MW00714270

Quality Improvement

A Systems Perspective

Quality Improvement

A Systems Perspective

William Roth

S^t^L

St. Lucie Press

Boca Raton London New York Washington, D.C.

Library of Congress Cataloging-in-Publication Data

Roth, William F.
Quality improvement : a systems perspective / by William Roth.
p. cm.
Includes bibliographical references and index.
ISBN 1-57444-236-8 (alk. paper)
1. Quality control. 2. System analysis. I. Title.
TS156.R673 1998
658.5′62—dc21

98-38861
CIP

International Standard Book Number 1-57444-236-8
Library of Congress Card Number 98-38861
Printed in the United States of America 1 2 3 4 5 6 7 8 9 0
Printed on acid-free paper

Biography

William Roth is a professor in the business department at Allentown College St. Francis de Sales. Along with Russell Ackoff, he is currently helping design the MBA program of the future at Villanova University.

Roth earned his bachelors degree in Geography from Dartmouth College, his masters degree in Social Work from the University of Pennsylvania, and his Ph.D in Management Sciences from The Wharton School. He currently has four books on management theory/quality in print. Three, including this one, were published by St. Lucie Press. He has published over forty articles on quality in leading national and international journals. Recently, Roth was asked by the Association for Quality and Participation to take the lead in designing a national program that provides comprehensive training in the systems approach to quality improvement.

Contents

Tables and Figures

Tables

Figures

Preface

Maybe a year ago some fellow systems practitioners suggested that I put together a book on the systems approach to quality improvement to be used in the classroom. They wanted a book that wades through all the buzzwords and hype and lays out the details of how to effectively start a comprehensive process in a simple, straightforward manner. They asked for this book because the systems approach, if properly introduced and supported, has never failed. They asked because the systems approach is generic and has been introduced successfully to a wide range of organizations in both the for-profit and the nonprofit sectors. They asked because the systems approach, by definition, is comprehensive, includes all the necessary pieces, but, more important, because it integrates these pieces in the necessary manner. Most of all, they asked because they had come to realize that the systems approach is the well-spring of all other approaches, of the quality improvement movement itself.

The question, then, of course, is, "If all of the above is true, why haven't more companies adopted the systems approach to quality improvement?" Actually, a growing number of organizations have. For the rest, there are a number of reasons why they have not, some obvious, some more subtle. The first reason is conceptual. Companies and most practitioners, including consultants, do not understand the true basics of quality improvement. They do not realize that it is nothing new but, rather, part of a progression at least several hundred years old. If they do not understand the basics or the roots of quality improvement, then they obviously have difficulty defining its requirements. As a result, they end up building and adopting models that are, for the most part, inadequate.

The second reason, on a more micro-level, is that, in general, we have worked hard to develop individual pieces of the quality puzzle, but have then stuck them together in a way that precludes the generation of a holistic perspective.

The third reason, also on the micro-level but having again to do with the basics, is that we continue to focus our efforts on the wrong thing. There is one primary focus to successful quality improvement efforts, just one. Everything else revolves around and feeds off it. If we miss that focus, or try to dilute it in any way, we are not going to produce the desired results.

The fourth reason is cultural. The systems approach to quality improvement, though proven, is in many ways threatening to our traditional management culture. It forces managers to truly make the changes we all have been talking about for at least ten years. Quite simply, in many cases, our managers are still not ready, or have not been trained properly, or some combination of these.

What we shall do in this book, therefore, is to present what might be considered an ideal model. It is the model toward which all others, including what has come to be called the Baldrige model and a similar model in the health care sector encouraged by the Joint Commission on Accreditation of Healthcare Organizations (JCAHO), are moving. Some companies will be able to adopt the systems model directly. The majority, however, will more likely use what we present as a standard to measure their current efforts against, improving and adding to or reshaping the pieces already in place.

With this objective in mind, *Quality Improvement: A Systems Perspective,* has three loosely-defined sections. The first digs back into history and explains the roots of the systems approach. It tells how the quality improvement movement fits into the historic whole, what the movement's macro-purpose is, and what its focus must be. It then lays out the systems approach to quality improvement in detail, covering all the nuts and bolts.

Section two takes us from the historic and theoretical into the real world. First it compares the systems approach to the Baldrige model, defining the strengths of each. Then it presents a detailed, *honest* case study of implementation of the systems approach in an industrial setting. It focuses on the problems that arose during this effort and how these problems were dealt with, either successfully or unsuccessfully.

The third section discusses major obstructions to the adaptation of the systems approach to quality improvement—the reticence of most top-level managers to change; the reticence of unions to adopt a new role; the reticence of academia to evolve—and explores the cultural origins of these obstructions. The book ends with an abreviated case study in the health care sector to demonstrate the systems model's generic qualities.

It is fairly well-guaranteed that by the time students finish *Quality Improvement: A Systems Perspective* they will know *what* should be done and *why.* It will then be left up to them to figure out the *how,* to discover what is possible in their own organizations and what is not.

Good reading, and good luck.

William Roth

1 Feeling Our Way

After Reading This Chapter You Should Know

- The two immediate producers of the quality improvement movement.
- The three approaches to dealing with the "operator" problem.
- The strengths and weaknesses of Frederick Taylor's scientific management.
- The origins of the human relations school of thought.
- The roles of self-directed work groups and labor-management councils.
- How the quality of working life movement differs from the quality improvement movement.
- How we insist on reinventing the wheel.
- Why empowerment is our major stumbling block.

Discovering Our Roots

The quality improvement movement began in the late 1970s and early 1980s. It was generated by the realization that the U.S. Achilles heel in the world marketplace was the quality and durability of its goods. The reviving European industrial sector and especially the Japanese had pinpointed this as their best chance to gain ground.

At first we paid little attention to their successes, convinced that we could not be caught. We had more capital to invest. We had more advanced technology at our disposal. We offered the best graduate-level education in business-related fields. In a relatively short period of time, however, that confidence faded as our competitors learned what we had to offer, then began improving on it, frequently with the help of experts such as W. Edwards Deming who, frustrated with the inertia he met in his own country, welcomed a receptive audience.

The U.S. began scrambling to maintain its lead. At this point, however, we ran into another realization, a common sense one that had been ignored since the beginning of the Industrial Revolution, since the beginning of our love affair with affluence and the things it made affordable. This second realization was that the resources upon which our abundance had been built were not inexhaustible. We could, indeed, run out of coal, iron ore, oil, good farm land, clean water, and forests if we continued to plunder our environment to feed factories. Planned obsolescence, the practice of designing products to last only a year or two so that new ones would have to be manufactured and bought as a means of keeping the economy healthy, was no longer an option.

A solution had to be found to both these problems, and that solution was quality improvement. It would help us continue to compete. It would help us to develop products with a longer life span. Quality had traditionally been the domain of the quality control inspector at the end of the process line who looked for defects, discarding or sending back products that did not meet specifications. This was obviously no longer sufficient. For one thing, it cost too much. Our competitors were now offering not only a quality product, but frequently at a lower price. We had to cut our costs. It was obvious that the answer to this challenge did not lie at the end of the production line where the damage had already been done, where time and materials had already been wasted. J. J. Juran made this clear. It was obvious that we had to back on up the line, and stop the mistakes from occurring in the first place.

Being basically a quantitative, engineering-oriented business culture, the first people we brought in were those with technical backgrounds—efficiency experts and operations research modelers. The challenge was to tighten up manufacturing systems and service processes until zero defects occurred, until no inefficiencies were possible. The employees involved, the "socio" part of socio-technical systems, according to the engineering school of thought, were secondary. Technology was primary. Employees serve the machines, the processes. Get the technology right, the employees would fall into place, and quality would improve.

Employees as the Sticking Point

We were very quick to find out that when the quest was for improved quality, this approach did not work. The operators turned out to be more important to the improvement process than expected. They did not cooperate. They were too difficult to get under control. They behaved in too human a manner, were too unpredictable, too likely to put their own needs before those of the machines and processes they served.

At this point, then, the focus in the exercise necessarily shifted to employees, to dealing with their lack of desire to change, no matter how obvious the organizational benefits of such change might be. Three ways of addressing this problem were defined, none of them new. The first was simply to get rid of the workers entirely. Then they would no longer be a problem. The inefficiencies left could be dealt with more easily. Automate completely. With our rapid advances in the field of computers this was a growing possibility.

Several companies actually tried this approach, but again it failed. Their realization was, first, total automation was extremely expensive; and, second, no matter how far you cut back on the workforce, there always had to be at least some employees involved, if only designers and maintenance people. A third realization was that such attempts to eliminate employees completely raised immediate and extremely serious social issues. Of course, the most important of these was, "What will we do with all the unemployed?"

This, again, was by no means a new question. It has been around since the beginning of civilization. In some primitive tribes the answer was simple. When a person could no longer contribute, was no longer productive, send him or her off into the wilderness to die. As we know, this sentiment continues to pop up even in the most advanced societies. It also continues to raise a more fundamental question which is, "What exactly should the purpose of the economic sector of a society be?" Should businesses function solely as vehicles which allow owners to make as much money as possible, to focus entirely on improving their own bottom line by continually eliminating employees. Should they be encouraged to do so? Will such a strategy in the long run pay off, even for them? Or must the needs of the other members of society be considered? And if they must, what is the most equitable means of distributing or redistributing the wealth generated? How can it be done so that the individual's incentive vital to the creation of society's wealth is not stifled in the first place?

The answer to this last question, the one about distribution and incentive, must be, "No one is sure." Several different schools of thought have developed over the ages. At one end of the spectrum we have the pure laissez-faire school which believes that it is all right, even necessary, to eliminate jobs in order to enhance efficiencies and increase the wealth of owners. The eventual result of this philosophy, however, of making untempered self-interest the driving force of the economic sector, has historically been severe disruption of the economy and of society; disruption that has led, in some cases, to revolution and a swing to the other philosophical extreme, to communism.

The economic philosophy of communism switches things around and makes the fortunes of the individual, the cultivation of individual potential totally secondary to the welfare of society as a whole. Carl Marx, perhaps the major spokesman for the communist economic philosophy, predicted that the free enterprise system would soon fail, drowning in the greed it unleashed, and that communism would prevail. He was wrong. While perhaps more idealistic, communism was based on false logic. Historically, it has been useful as a vehicle for facilitating the rapid, but often bloody redistribution of existing wealth. After that, however, the citizens laboring under it have quickly became dissatisfied with the lack of opportunity to benefit directly from their own efforts.

A more moderate alternative had to be found, one that incorporated the best of both extremes. A synthesis had to occur, and that synthesis had to be based on the realization that the purpose of the economic sector in any progressive society should be *to improve the fortunes of individuals as part of society.* The flip side of the coin is that its purpose should *be to improve the fortunes of society as a whole in a way which, at the same time, most strongly encourages the development of individual potential.*

The economic sector helps to meet these challenges. It does so, mainly, by providing goods and by generating the wealth to purchase them. Both the goods and the wealth are gained by people through work, usually for an employer. If people are not allowed to work, if, for example, they are replaced by machinery, or if they are paid only a survival wage, although the goods might still be generated, the wealth produced will be funneled to a very small, powerful elite, and the goods will rapidly become unaffordable to most.

Henry Ford in the early 1900s understood this logic. He showed, despite the loud protests and threats of a majority of his fellow industrial leaders, that the best way to improve his own fortunes *long-term* was to pay employees enough so that they could afford to buy the cars they were manufacturing, thus improving *their* purchasing power and fortunes as well.

A New Tack

The second alternative when dealing with an uncooperative workforce was to focus on training and to condition employees to function as parts of the machines they were running. “Leave your human characteristics at the door when you come to work. Turn into a piece of machinery or a counting process that does not tire, does not have to go to the bathroom, and does not become distracted worrying about the daughter's low grade in arithmetic."

This approach was not new either. In fact, it had been extremely popular since the middle of the Industrial Revolution. Adam Smith, a Scottish teacher and philosopher, the father of laissez-faire economics, talked in the 1700s about making jobs as simple and repetitive as possible as a means of increasing productivity and of enabling the unskilled to find employment in the new "manufactories."

In more modern times, Frederick Taylor, an engineer who did his work mainly in Philadelphia and the Lehigh Valley of Pennsylvania, developed the concept of "Scientific Management" by applying the principles of science to workplace management. Science in the early 1900s was basically empirical, which means that nothing scientists could not quantify was taken into consideration. Taylor tried to run the workplace by the numbers, quantifying everything, including workers and their activities. He defined every part of the operation in terms of efficiencies. He said that workers should be as dumb as possible, should be encouraged to think as little as possible because thinking wasted time. He advocated the use of time-motion studies for every job to identify the procedure that produced the greatest output with the least amount of exertion. He then advocated tying the employee's pay to their productivity.

Attempts to turn workers into machine parts, however, also failed because people and machines suffer at least one critical difference. As Ackoff and Emery say in their book, *On Purposeful Systems,* while machines are "purposive," that is, given their purpose by the designer, humans are "purposeful." People have needs, wants, and values of their own. They work in order to meet these. If owners do not take the purpose of employees into consideration, along with the purpose of the technology and the purpose of the organization, the workers are not going to be as productive. Perhaps they are not even going to cooperate.

Finally on Course

Frederick Taylor made an important contribution, one which continues to help shape the modern day workplace. But, again, it needed to be tempered, to be humanized. The third alternative, the only one left at this point, was to focus on the needs of employees as well as those of technology. This was a whole new ball game, at least in terms of improving quality. A group of management theorists labeled the "Human Relations" School appeared during the late 1800s and early 1900s as a reaction to the engineering approach, as a reaction to the mechanization of employees, to Taylor's scientific management.

Leaders of the Human Relations School include Follett, Walton, Mayo, and Roethlisberger as a team, and Herzberg, McGregor, and Argyris. Mary Follett, an American economist, said that workers should become more involved in making decisions that affected them, that boundaries between workers and managers should be broken down, as well as those between units. She also urged society to begin realizing that employees do not leave their needs at the door when they come to work and that they have the same needs on the job as they do when at home. She said that productivity would increase if management strove to satisfy these very human needs.

Oliver Sheldon, an English businessman, agreed that workers were more important than machines and that the primary purpose of business was not so much to make money as it was to meet the needs of the community served, that community included employees. Sheldon also said that workers should help decide the conditions of work.

Elton Mayo and Fritz Roethlisberger from Harvard "slept with the enemy," so to speak. They used scientific research to substantiate Follett's and Sheldon's claim that making satisfaction of employee needs the primary concern would increase productivity. In 1927, at the Western Electric Plant in Hawthorne, MI, they ran a series of experiments to test the effects of work environment improvements on productivity. The most important result of these projects was the realization that the employees reacted positively in terms of productivity to any interest in their situation shown by management, whether it actually improved things or not. This result was labeled "the Hawthorne effect."

Frederick Herzberg, another social scientist, differentiated between *satisfiers,* factors in the work environment which generally had a positive effect on employee productivity, and *dissatisfiers,* factors which generally had a negative effect. In his studies, the satisfiers included achievement, recognition, the work itself, being given responsibility, and advancement. The dissatisfiers included company policy and administration, technical supervision, interpersonal relations with supervisors, salary, and working conditions.

Chris Argyris of Yale showed that employee attitudes are shaped by the way managers interact with them. If managers treated employees like children, they behave childishly. If managers treat employees like responsible adults, they tend to respond like responsible adults.

Douglas McGregor of MIT produced Theory *X* and Theory *Y* to differentiate between the two dominant philosophies of management. Theory *X* said basically that workers wanted to do as little as possible and needed to be driven and closely controlled by management. Theory *Y* assumed that workers have a more positive attitude and needed only to be encouraged and guided. It said that workers wanted to perform well and to use their potential effectively in terms of improving productivity. It said that the problem was not the workers, but the managers; that managers, rather than driving the workforce, had to change their own attitudes, to become more positive and supportive, if they wanted the workers to become more productive.

Going All the Way

Another major contributor was Eric Trist of the Tavistock Institute of Human Behavior in London. In the coal mines of England, during a research project, he discovered an instance where the human relations approach had been carried to its logical extreme. In one of the mines the workers had actually taken total control of the operation. There were no supervisors. The miners decided who would work when and as part of what shift. They trained each other. They organized the actual work. They took charge of discipline and hiring. They had truly been empowered. As a result, productivity had increased greatly.

This new autonomous work group approach (currently called the self-directed work group) was tried in several other instances including a textile mill in India and a motor parts company in England. In every case, however, it was opposed strongly by management, both corporate and union. In almost every instance it was not allowed to spread and was eventually shut down.

The one major exception was the Scandinavian countries. The new management model was introduced here by Eric Trist and Fred Emery, also of the Tavistock Institute, at the invitation of the Norwegian government. While successful at the pilot sites, it did not spread as rapidly in Norway as it did in neighboring Sweden where efforts at the Volvo and Saab automobile manufacturing plants have gained world-wide attention, and where thousands of similar employee empowerment projects have been mounted in all sectors.

The key to success in Scandinavia was most certainly cultural. Scandinavian countries are socialistic. The economic system of socialism is an attempt to find acceptable middle ground between the laissez-faire and communistic schools. The espoused purpose of government in Sweden is to provide a decent quality of life for all citizens and to encourage the development of individual potential. Everyone is guaranteed employment, even if it has to be government subsidized. Taxation is heavy in order to support a multitude of social programs, but at the same time, owing to the degree of *physical* and *emotional* security enjoyed, the *scarcity mentality* is not as strong here as in other cultures. The scarcity mentality stems from the belief that we do not have enough wealth to insure security for all, to meet everyone's needs and desires. As a result, we must battle constantly to gain more. Because the scarcity mentality is not as strong in Scandinavian countries, the workplace is not as competitive. Employees on all levels are more likely to work together for the common good.

A second vehicle which allowed employees to help shape their work lives evolved independently in Eastern European communistic countries during the 1920s and 1930s. It was called the workers council (currently known as a labor-management committee). Such councils were composed of representatives of labor and of upper-level management. They met monthly. The purpose of these meetings was to forge a direct link between the workers and executives. Through the workers council, labor was able to learn about and to help shape the organization's objectives and future. Labor's representatives were also able to bring issues directly to the top without first wading through layers of hierarchy. Workers councils spread throughout Europe and are now required by law in a number of countries.

Quantities vs. Quality

Most members of the Human Relations School, most of the projects we have discussed, were part of what was called the quality of work life movement. We must realize at this point in our discussion, however, that their focus was not on improving the quality of the product. Rather, it was on finding ways to improve the work environment and to meet the employee's personal as well as job-related needs in order to increase the quantities of products generated.

It is only during the last 20 years or so that our attention has shifted more in the direction of product quality. It has been only in the last ten years or so that we have begun to understand the linkage between the quality of work life movement and the quality improvement movement, that product quality cannot be improved without also taking into consideration workers needs and the quality of the work environment.

Unfortunately, U.S. industrial leaders in the newer quality improvement movement have, too frequently, treated the challenge as an entirely new and personal one, largely ignoring the lessons of history, largely ignoring the research that had been carried on and the conclusions drawn during the first part of the century. We have also ignored what is going on in Scandinavia with autonomous work groups and in other European countries with workers councils. We have been determined to come up with our own model, one which best fit our own interpretation of our workplace reality.

In the early days of the quality improvement movement, for example, it was not unusual to hear executives kicking off quality efforts to direct employees to focus on product and process improvements solely and to say that work environment improvements were not to be taken into consideration. Their fear, of course, was that team members would focus on work environment issues exclusively and end up costing the company lots of money while doing nothing to enhance the bottom line.

The one country we did pay attention to, strangely enough, was Japan. This was probably because W. Edwards Deming had taken the lead there and because we felt threatened by the rapid advances being made. Still, however, we were very selective about what we accepted from the Japanese model. Dr. Deming, after his return to the United States, was extremely frustrated by our preoccupation with the quantitative tools he had developed to the exclusion of the rest of what he had taught the Japanese and what he had learned from them. He considered the quantitative tools to be less than 10 percent of the necessary whole.

Things Become Real

During our early efforts and frustrations, however, as we have worked to reinvent the wheel, we have also eventually learned something very important. We have learned that a process paradigm much more sophisticated than anything currently in place, at least in this country, was needed; one which went far beyond the work of the Human Relations School, and certainly beyond that of the engineering school, one which effectively combined the strengths of the two.

The question was how should we start developing this paradigm, especially when time was short, when the pressure was on, and when other countries were steadily eroding our lead in the world marketplace. We began in the only way we knew. After the efficiency expert approach failed, we adopted a training-centered approach. We brought in more experts, this time those whose specialty was human motivation. Their job was to train employees to care about quality and to give them the interpersonal skills necessary to improve it.

This approach worked to a degree, probably because Mayo's and Roethlisberger's "Hawthorne effect" kicked in. Still, however, the desired results in terms of quality improvement were not being realized, and we grudgingly began broadening our perspective. People started talking about the need for cultural change and for empowerment. This was also when companies began trying to piece together comprehensive organization-wide efforts.

Quality improvement quickly became the buzzword in corporate circles. One immediate result was that consultants flocked to the field. The market was soon flooded with experts and gurus. They came from all directions. Some had academic backgrounds in psychology, organization behavior, or organization development and provided mainly training for team members. Many were efficiency experts, offering the same skills they had offered before, only packaged differently. A great number were ex-employees of organizations who had some practical experience with a quality process and were passing on what they had learned. And finally, too, many were opportunists and marketers who patched pieces together and won contracts with their presentation skills rather than because of what they offered.

The problem was that no credentials were required of the consultants. None had been defined. There was no serious training available for those who wanted to teach quality improvement skills, much less to implement a quality improvement process. There were lots of seminars. Colleges offered courses, but nothing comprehensive had, as of yet, materialized. Pieces of the puzzle were available, but they had not yet been fitted together properly. This was partially owing to the fact that nobody was actually sure of what the pieces to this new whole were. As a result, anyone who published a book, or an article; anyone who had experienced a process, or, at worst, had memorized all the right buzzwords, could offer a model and claim to be an expert.

The Challenge

Meanwhile, back at the corporation, developing an organization-wide effort to empower employees was something new to most. Organization-wide, rather than departmental or divisional, was not the standard approach, and the concept of empowerment was not only non-traditional but also frightening to many. So companies did what they usually do when starting a new, unfamiliar, possibly threatening project. They picked a "rising star" to head the effort: someone who was hard working and dependable, someone who was locked in, who understood the rules of the game, someone who they were sure would not do anything unexpected. They then guaranteed upper-level management support, gave this person a fairly sizable budget and a staff, and said, "It's all yours. Make it work, but don't change anything, at least anything that will affect the way we are currently running the company."

Most such efforts obviously failed. They failed because, under the circumstances, with the traditional turf battles still going on and with everyone protecting his or her own turf, it was impossible to successfully mount an organization-wide effort and because, within the current management culture, it was impossible to truly empower participants. Today, organization-wide integration and real empowerment are still missing in most cases. While a number of organizations have made some progress toward the first requirement, very few have even truly acknowledged the second. In terms of real employee empowerment, it is interesting to note that the few modern-day U.S. companies that have actually fostered it refuse to attribute their decision to a formal quality improvement effort. Most do not even have a formal quality improvement effort in place. We are talking about the Harley-Davidsons, the Lincoln Electrics, the W. L. Gore Corporations, and the Johnsonville Foods Companies.

It is also interesting to note that when real empowerment has actually cropped up as a result of quality improvement efforts, it has usually been shut down. This was the case of the much analyzed Gaines dogfood plant effort in Topeka, KS, of the General Motors parts plant in Bay City, MI, Tom Peters talks about; of the ATT plant in Reading, PA, and of the International Paper mill effort which we will discuss later. Real empowerment has materialized in these instances, but has been isolated rather than allowed to spread, and eventually has been smothered because it conflicted too greatly with the organization's management paradigm.

Real empowerment has materialized in the above instances, invariably, because one person somewhere at the top has decided to let it materialize. It has dissolved, almost inevitably, when that person left the position, either by choice or, more frequently, by corporate decision. Real empowerment has not been encouraged by a majority of those controlling the fate of our economic sector. Criteria for the much valued Malcolm Baldrige Award, the United State's highest award for quality improvement, for example, encourage only partial empowerment. Another example would be the current JCAHO guidelines in our turbulent health care sector which fail to properly encourage the real empowerment of a highly educated, highly skilled workforce.

The basic problem, the thing most responsible for these ongoing shortcomings in our usually sincere attempts to change is, as we have said, the nature of our workplace culture. To empower means to trust. To empower means to be willing to turn over authority as well as responsibility. To empower means to be willing to openly share information. Our workplace culture is too competitive. Such thing cannot happen.

While other industrial powers have realized the value of finding ways to encourage employees to work together, to integrate their talents, to view each other as members of one team competing against other companies and other economic powers; we are still fighting it out in the in-house level, driven by the misguided belief that this will allow the cream to rise to the top. Perhaps Peter Drucker put it best when he said that the fiercest competition (conflict?) in the U.S. economic sector is not between companies, but within individual companies, and that it is also far less ethical.

The point is that until we get beyond this "old boy" attitude, we are going to have trouble mounting successful quality improvement efforts. Our traditional evaluation and reward systems, for example, are purposely organized to stimulate competition (conflict?) They alone preclude real integration and empowerment from occurring. We need a new paradigm. Which brings us to the second key ingredient necessary to successful cultural change and necessary to a successful quality improvement effort.

This second ingredient is a comprehensive approach that is truly participative, organization-wide, integrated, ongoing, and that allows for constant feedback and learning. What we are talking about, of course, is a systemic approach. Very few companies have one. Very few quality consultants and practitioners even know what the systems approach is. Without the systems approach, however, chances of real success are extremely slim to non-existent. A systems approach, by its nature, guarantees the necessary empowerment. At the same time, it allows employees to overcome the obstacles bred by a competitive environment. The systems approach, whether introduced knowingly or unknowingly, is the only way, ultimately, to generate the necessary cultural change.

But what, exactly, is the systems approach? That is what we will talk about next.

Topics for Discussion

1. Why do you think your organization really began its quality improvement efforts?
2. Which of the three approaches to the "operator" problem or what combination thereof does your company use?
3. What were Frederick Taylor's major contributions with scientific management? What factors did he overlook?

4. Who were the key figures in the human relations movement and what were their contributions?
5. Could autonomous (self-directed) work groups operate effectively in your organization?
6. Why do most U.S. companies want to "do it their own way?"
7. To what degree has your organization empowered employees?

2 The Systems Approach

After Reading This Chapter You Should Know

- The steps usually taken in starting a quality improvement process.
- The dangers of combining downsizing with efforts to improve quality.
- How organizations usually choose their quality improvement model.
- When most quality improvement efforts begin to fall apart.
- The basics of systems theory.

The Learning Curve

Quality has become a central issue in our business community. Most companies of any size now have quality improvement efforts in place, no matter what they might call them. If these companies have followed the pack in terms of designing their efforts, however, chances of long-term and frequently even short-term success are limited. A cycle that is not only non-productive in terms of the desired results but can actually be counter-productive has been set in motion.

Before getting into the systems approach as a solution, let us move through the steps usually followed when a corporation mounts a quality improvement process. To begin with, the CEO or head of a major unit, having read enough or heard enough to believe, and rightfully so, that quality is a key to the improvement of the company's as well as his or her own fortunes, decides it is time to get something started. That person discusses the decision with key reports, then either staffs a quality department or assigns the quality function to an already existing department.

The next step, as we have said, is to choose a head for the new function. Companies lacking real commitment or with a lot to learn assign this job to someone as an additional responsibility. In primary industry that person is frequently an engineer. In the service sector, it is a member of the human resources department. In companies that want to produce meaningful results, the position is full-time and reports directly to the CEO. The new quality process head is given several months to prepare. He or she begins by reading as much as possible on the subject, attending seminars, and visiting other companies with processes in place. The quest is for a model, if the CEO has not already picked one, as a way of organizing the effort.

The new process head reports back frequently to ensure that the boss is still supportive and understands that progress is being made. The clock is ticking. The company has invested a lot of money in this effort and is looking for results. Also, for the process head this a fairly risky gamble. Being the leader of a quality improvement effort is not a step in a traditional career path. It provides tremendous exposure and could allow the individual to jump ahead, but it could also spell disaster. Everything depends on the results produced in an almost entirely new arena.

First Big Mistake

About this time we get the first serious indication of the company's true intentions. One of the strongest clues that a CEO's major concern is short-term bottom line improvement rather than long-term quality improvement is a decision to downsize, to get rid of deadwood. Perhaps organizations making this decision believe that having fewer employees to train will ease implementation. Perhaps they believe that a little fear, proof that everyone is expendable, will help motivate. Perhaps they plan to fund the quality effort with the money saved by downsizing. Or perhaps they feel that with change in the air, this is the ideal time to make such a move.

Improved quality, they believe, depends on creating a leaner, meaner business machine. The up-front need is to decrease the number of management layers in order to improve communication and to speed up the decision-making process. The need is to eliminate duplication of responsibility and function. It is to get rid of bureaucratic slowdowns by contracting traditional staff functions out to companies that specialize in those areas. According to their way of thinking, all of the above can be accomplished most efficiently by eliminating employees.

Whatever the logic, such a move is self-defeating at best, disastrous at worst. It is also illogical if enhanced product or service quality is an objective.

The ingredient that is key to such efforts is the same ingredient that was key to earlier productivity increasing efforts—employee commitment. The amount of evidence that has built up through the years in support of this statement is irrefutable. Improvements in technology are important, yes, but if the employees controlling the technology are not on board, the desired results are rarely achieved.

The number one goal of any quality effort must be to gain employee commitment. This is done by showing employees respect, as the old human relations school teachers preached and as a growing number of new world business leaders have proven. When a company announces that it wants everyone to contribute to the improvement of quality, to become part of the team, and in the next breath it adds that as a first step in this direction it is going to lay off 10 percent of the work force, the quest for commitment is, to say the least, crippled and cannot be revived according to published estimates for at least three years.

Downsizing breaks lines of communication, especially between levels. The rumor mill dominates. Anything that comes from top-level management is automatically suspect. The quality process, for example, is viewed by many actually as a cover for an on-going efficiency study, an attempt to find ways to cut more jobs. Information is hoarded in order to enhance one's value when the free dissemination of information is critical to quality improvement. Problem-solving efforts lose integration. Units focus on their own problems, worrying less about the effects their decisions might have on the fortunes of other units. Survivors of the downsizing are forced to take on extra duties without proper training. Discipline either becomes more severe, or disappears entirely owing to management's desire to regain the confidence of employees. Everyone ends up working longer hours for the same salary or for an increase so small that it is insulting. Job security becomes an on-going issue no matter what management promises.

In essence, when a downsizing occurs, the workload goes up, training and moral suffer, and, to make things worse, the anticipated bottom-line improvement is short-lived. Suspicions soon arise that the exercise will eventually have to be repeated.

Why We Downsize

Downsizing is obviously *not* a way for management to show employees that it respects them. Yet the tactic continues to be coupled with corporate quality improvement efforts. It is as if on one side of the scorecard there are columns of numbers that need to be juggled and on the other side there are employees with feeling and needs. It is as if management has trouble understanding the connection between the two sides, does not want to understand the connection between the two sides, and believes it can do things on one side that will not be noticed on the other.

The problem might be one of stakeholder prioritization. Stakeholder is a systems term. It refers to groups whose lives are affected by an organization's actions. Key stakeholder groups in the business sector are usually the organization's owners (shareholders), customers, suppliers, employees, and the community. The most important of these groups is considered by most to be the owners, represented by Wall Street and the financial community. Wall Street, however, is interested mainly, or, perhaps, only in short-term improvement of the bottom line, in immediate gain for its clients. From this perspective, downsizing makes good sense because it produces quick results.

If quality improvement is the objective, however, Wall Street's perspective is off-target. Thus, the financial community cannot be considered the most important stakeholder group. The second choice, popular amongst companies with relatively mature quality efforts, is the customer, both internal and external. If everyone works to meet customer needs, improved quality will be assured. But this choice is also off-target. Customer satisfaction is a measure of how well the effort is doing, not the driving force behind it. Customers cannot make a process succeed. Therefore, customers cannot be the most important stakeholder group, either.

Which, then, is the most critical stakeholder group? To find this out, just ask yourself a simple question. "Which of the stakeholder groups is most capable of changing the attitudes of the others?" If the investors are unhappy, which group can make them happy by improving sales and the bottom line? If the customers are unhappy, which group is best qualified to find out why and to make the necessary changes? If suppliers are unhappy, which group can deal with the issues involved? If community leaders are upset, representatives of which group need to sit down with them and talk?

The answer is obvious. Only one group is tied in with and can positively influence all the others. This stakeholder group is the employees. Much of top-level management has been avoiding this realization, despite its obviousness, for a long time now. Until they face it, make it the core of their approach to improving quality, their chances of success will remain limited.

Back on Track?

The generation of employee commitment to the process is key. Let us say that our new quality department head realizes this and that the company has not downsized, but has, instead, promised to do everything possible to avoid downsizing if the employees will pitch in. What happens next?

The realization comes fairly quickly to the new head of quality improvement that, while the theory part might not seem so difficult, implementation is another matter. How do you develop the necessary level of understanding in a workforce of thousands? How do you even attract the attention of employees when most of them are already overloaded with work? Where in the organization do you start? How do you change management practices that have been in place and successful for 20 years? What kind of training is best? Which of the hundreds of brochures received from consultants since word got out makes the most sense?

It was not unusual for panic to set in as the months passed and the quality process head begins realizing how much has to be learned and how much complexity there is to changing the culture of any organization. Eventually, that person is forced to choose a model or to become enthusiastic about the one the CEO has chosen. The winner in the first case is usually the model that seems safest. Other companies have vouched for it. All the proper sales materials have been presented. Things are to be introduced in a way top-level management is accustomed to. And it is expensive enough to be taken seriously.

Once the choice is made, the quality process head works hard to sell it to the rest of the workforce although, because so little of the necessary homework has been done, he or she is frequently not sure of what it is going to produce. The frantic public relations effort that follows is facilitated by the fact that the first step in almost every consulting firm's package is extremely straightforward and produces extremely visible results. It is to familiarize employees with the need for improved quality, providing an accurate definition of quality and an understanding of what quality improvement entails, and then to train them. Groups of managers and office or shop workers are either shipped away to a conference center or invited into classrooms where they learn what the different slices of the quality pie represent, how their culture is necessarily going to change, and then trains them in team building, group process, problem identification and solution, conflict resolution, and creativity enhancing techniques.

During this phase the quality department head receives the first quantifiable and displayable indicators of success to carry back to the boss. The growing number of sessions held, the number of employees who have been familiarized and trained, as well as the growing number of quality-related posters, notebooks, quotations, pins, and paperweights that have been distributed, can be pointed to as proof of progress. In large companies hundreds, even thousands of employees are trained up-front. Those who are chosen to head the process in their individual units, usually managers, receive additional hours, even days, of training.

This first phase usually takes months, even years. During that time, the executives also attended sessions to become familiar with the model and with their role in the process. During these sessions the executives frequently begin identifying projects that they want their employees to work on.

Hitting the Wall

Eventually, however, the quality process head has to begin worrying again. The next step is to transfer what has supposedly been learned in the conference centers and the classroom to the workplace, and no one is quite sure of how to do this. Teams are the obvious vehicle; but as the team building phase begins, things do not work the way classroom notes say they are supposed to.

At that point, one of two things can happen. The first is that the executive group takes charge and assigns the projects it has come up with to lower-level teams that it has formed. These teams are composed mainly of managers, or at least are headed by a manager. They also, however, include at least some lower-level employees. The second possibility is that the process head takes charge of putting teams into place, the people who have received additional training at the seminars being responsible for getting things started in their units.

Historically, the model we have worked from in team building is the Japanese quality circle, but, again, there are cultural differences to be considered. In the Japanese workplace a sense of respect and a tradition of teamwork is found that does not exist in our workplaces. Some U.S. organizations try quality circles for a while, but quickly discard them and, again, begin designing their own vehicles to suit their own culture.

The team building phase is the point at which many quality improvement processes begin to crumble. The teams do not meet on schedule. Members do not show up, or are not being allowed to show up because of their workload; or team members do show up, but nothing gets accomplished; or team members show up and spend their time arguing; or, though the manager leading the team has agreed to "take off his hat" during meetings and to share the stage, to share decision-making authority, it does not happen.

At this point, the original quality process head frequently decides it is time to move on. He or she does so, declaring that progress had been made but that, as we all know, it takes years for quality improvement efforts to bear fruit. A successor is chosen, and the first thing the new head usually does upon taking over, in order to make his or her mark, is to reshape the pieces in the organization effort, to bring in a new team of consultants, a new model; or to decide that a need exists for additional training.

And so the process moves slowly forward until at least part of the desired change occurs, or until upper-level management loses interest and moves on to the next buzzword (reengineering?), or until the effort becomes too expensive and a manager with a financial background is put in charge with orders to begin cutting costs.

Meanwhile, Down in the Trenches

The story, however, does not end here. Especially in larger companies, some unit heads, eager to make an impression, have not been willing to wait through the long start-up and training phase. They have done their own reading and have begun organizing their own, individualized efforts. Upon hearing about this challenge, the quality process leader might turn to the CEO to block it. The CEO, however, is not usually willing to stop such activities. The Board of Directors is currently stressing the need to decentralize. Emphasis is on giving as much autonomy as possible to the major corporate units, and the CEO does not want to set an undesirable precedent.

Another reason units strike off on their own is that the quality department, with a staff of maybe five or six, has made the mistake of trying to organize the entire corporation at once, instead of developing pilot sites that can be used as training grounds and spreading slowly out from them. When the managers who have received the additional training arrive back to their units with orders to get things started, even if, by some miracle, they have learned all they need to know, their chances of accomplishing anything of value are slim. Most are quickly overloaded with other operating responsibilities and do not have the time. Even if they do have the time, when they start building teams, a lot of questions arise. Corporate quality staff has promised to help answer these questions, but because it has so much ground to cover, members are hard to contact.

So the managers in charge improvise. Generally, such improvisations fail and they end up simply going through the motions with no hope for real results. Sometimes, however, they come up with a model that works and begin producing positive change.

No matter what the situation, allowing units to develop their own, individualized approaches can be a very serious mistake. Too often, the result is a hodgepodge of fiercely defended, poorly integrated, partially developed efforts. Units are unable or unwilling to learn from each other and sometimes refuse even to communicate or to share information in their effort to win the competition, to make the best impression. One approach to quality improvement should be chosen. The CEO should mandate that it will be implemented organization-wide and that any modifications to this approach must be approved by the head of quality or by the CEO, or by both.

So What Is Missing?

We already know the answer to that one. What is missing in almost every case including that of Baldrige Award winners is real empowerment and a comprehensive, systemic model. We have mentioned this concept of systems several times in relation to successful quality improvement efforts. Now it is time to focus on it and to explain what we are talking about. Quite simply, what we are saying, again, is that the most important underlying reason for the continuing failure of quality improvement processes is a lack of the necessary systems framework.

We see looks of surprise and skepticism. "But what," you ask doubtfully, "does the systems approach, a trend that came and went at least 15 years ago without receiving that much attention, have to do with TQM?" The answer, quite simply, is that the latter is essentially a clone of the former. Only the name has changed. To put it more exactly, TQM is nothing more, nothing less than the systems approach to management, perhaps in more comfortable clothing, but the systems approach, nevertheless.

"If that's so," you ask, still doubtful, "why has it not been pointed out in the literature or by gurus?" The answer is that it has, by the real gurus, the Demings and Ackoffs, and Jurans. We just have not been listening, or we have been listening selectively. Part of the reason we have not been listening is that very few of us understand what the systems approach is about. We never did. We did not 15 years ago, we do not now. In terms of quality improvement, then, it follows that if we have not yet succeeded in grasping the original, foundational concept, we are going to have an extremely difficult time understanding the modern-day version, the clone. We are going to have a difficult time understanding the ingredients and interactions necessary to successful quality improvement.

Proof of this suspicion lies in the way most companies continue to mount their quality efforts. They do so, as we have said, without completing the necessary homework first. They do not complete their homework because they are in a hurry, or perhaps because they do not really understand how much homework is involved. The result is that most efforts begin in a piecemeal and frequently disjointed manner. The familiarization, planning, team building, training, measurement, and reward pieces are frequently not well thought out. They are thrown together, they are poorly shaped. Rarely, if ever, are they integrated to the necessary extent.

What companies end up with, then, in systems terminology, is an aggregate of pieces rather than the well-integrated, interdependent whole when such a whole is the essence of the systems approach and, therefore, the essence of successful quality improvement.

What, Exactly, Is the Systems Approach?

It is time now to get specific. What exactly is this systems approach? Actually, in theory at least, it is quite easy to understand. The systems approach, for one thing, is an attempt to wed the previously discussed engineering and human relations schools of management thought, but it goes way beyond them. The systems approach's origins are found, surprisingly enough, in the biological sciences. A man named Ludwig Von Bertalanffy, in the 1940s, made an announcement based on his research in botany that shocked the scientific world and eventually affected every realm of scientific investigation. Von Bertalanffy's discovery, his simple, foundational statement that changed everything, was that *a whole is more than the sum of its parts.*

What he meant was that any system, any whole, or group of parts dedicated to the same purpose, be it a healthy plant, a human being, a efficient machine, or a successful organization, *possesses critical properties that none of its parts possess, properties that arise from the interactions of its parts and give the whole its identity and character.*

When we dissect an organization, then, or a process, when we break it down into its parts, these critical properties are lost. The whole loses its unique identity and becomes, instead, simply an aggregate, or a collection of pieces with characteristics that can be added together, but that do not include the critical, unique ones which make the whole distinctive.

The loss of these critical, unique characteristics is what occurs when we analyze. Analysis is the act of breaking an organization or process down so that it is easier to study. We do so because we believe that a thorough understanding of the parts, of the divisions, departments, and work units will lead to the desired understanding of the whole. Sounds logical, but we are usually wrong. Analytical thinking has worked pretty well with mechanical and frequently with innate natural systems. But when a human or social (as opposed to technical) system is involved, our chances of success diminish greatly.

The next question, then, must be, "Why are human or social systems so different?" The answer is that while the parts of mechanical and innate natural systems are stable and, as a result, are good subject matter for a slow, methodical study, some parts of social systems are constantly changing and other parts are not. What is true about them one moment may not be true the next. This makes it extremely difficult to assemble the characteristics of social system parts into a composite that makes sense. One is saddled with the never-ending job of having to go back to fine-tune, adjust, or reorganize.

The reason for this volatility in social systems is, again, quite obvious. At the same time it has, again, escaped most of us. This is why it has escaped us. While the parts of mechanical systems have one set purpose programmed into them that cannot be altered unless the programmer decides to do so, the parts of organization social systems, as we have said earlier, have more than one set purpose. They have the organization's purpose, but they also have the purpose of work groups and the collective purposes of the individuals who run the organization which can change momentarily, and which are usually more important to the part than to the organization.

How Does the Systems Approach Work?

The systems approach understands this difficulty and tries to deal with it realistically. It does so by addressing the definition of purpose in a non-analytical manner. It forces us to think synthetically, to first look outward, instead of inward, to focus initially on the role of the organization in terms of its containing environment. It forces us to try to define a purpose which is common to all social and technical system parts in terms of this environment, one which is general enough to be universally accepted. The product of such an exercise is the vision or mission statement with which most of us are familiar. Such statements are one of the few remaining survivors of the systems period.

Once the organization's role in terms of the containing environment is understood and agreed to, the pieces can then begin defining their relationship with each other in terms of that role. This is where most organizations run into trouble. While the mission statement might be systemic in nature, organization culture is not. When the two clash, the organization culture usually prevails. Focus remains on the work unit as the center-point of employee reality, rather than on the organization as a whole. As a result, the vision statement becomes more of a showpiece than a cornerstone.

In order to further clarify the difference between the analytical and systems approaches when applied to organization improvement, the analytical approach, once again, focuses on the pieces. It says, "Our first objective is to try to figure out how to get Harry and Mary, or production and marketing to work together. I will talk to production and find out what their concerns are. You talk to marketing. Then we will compare notes and try to come up with a better way of doing things. After that, we will try to add product development to the team."

"Eventually, when everyone in the organization has been fitted together in an effective manner, we will explain our overall objectives to them. But we do not want to do that up-front. We will wait. We do not want to distract the individuals and groups we are working with. We do not want to draw attention away from their initial challenge."

The systems approach, on the other hand, says, "Our first objective is to define our organization in terms of the market we are trying to capture, in terms of the culture we are part of, in terms of our suppliers, in terms of our workforce as people rather than as machine parts, in terms of our owners. We will begin by putting together an overall organization purpose or mission statement acceptable to all of these key stakeholders.

"Once we have identified this mission, we will start identifying the characteristics our organization needs to fulfill it, again with everyone's input so that the results are richer and so that commitment is generated. Next we will get down to details. We will address the question of how we should actually design the products, the manufacturing processes, the management systems, and the work environment so that they possess these characteristics."

With the systems approach, it is only after these design details are in place that we start working with Harry and Mary or production and marketing, trying to help them fit comfortably into the model created. But more often than not this latter step has become unnecessary. By being involved in the exercise from its inception, Harry and Mary or production and marketing have gained ownership and have already begun making the changes in their attitudes and their interactions necessary for success.

Another Key Difference

Another key difference between the analytical and systems approaches is the way in which systems thinkers define organizations. They do so in terms of three macro-systems, each one of which permeates and connects every part of the organization. The three macro-systems are function, process, and structure. The three cannot be separated. They are more entwined than Alexander the Great's Gordian Knot. Each category shapes and is shaped by the others. Each is indispensable to the others. Together, they are the glue that holds the organization together. As a result of this relationship, they cannot be segmented, but must be dealt with as a whole.

Function focuses on the organization's role in the larger system of which it is a part. Function has to do with the mission/vision discussed earlier. We define the purpose of each division or department in terms of the organization's mission. We then define the mission of each section in terms of the department's mission. This process continues on downward and outward until the individual employee is reached.

Process has to do with how an organization meets its mission or fulfills its function on all levels. It has to do with the way widgets are made, the way information is gathered and distributed, the way messages are delivered, decisions are arrived at, rewards are defined, and developmental opportunities are allocated. In systems terminology, the process category has to do with how socio-technical systems, as defined by Eric Trist and his associates, are organized and integrated in order to achieve the best possible results. Its focus is on generating the organization characteristics identified during the definition of the organization's function.

Structure also has to do with the realization of these characteristics, but the less fluid ones. It has to do with how the pieces should actually look and fit together on an organization or flow chart, how the sections, departments, divisions should be organized, who should report to whom on a solid and dotted line basis, and how the machinery on the floor should actually be laid out.

One More Difference

Another difference between the systems or synthetic approach and the analytical approach to organization change is the former's shift toward effectiveness as a measure of success and away from efficiency. Those still practicing the analytical, break-it-down approach seek the most efficient model, the most efficient solution. Efficiency is their sole objective. Those still practicing the analytical approach continue to think mechanistically. They treat the organization, including employees, as a machine, as a collection of parts, the performance of which must be optimized.

With machines, when we are trying to make improvements, we define things quantitatively, numerically—the size of a gear, the number of amps required, the number of hours before a part wears out, or the number of workers it takes to reach a certain output.

When we focus on the interaction between parts rather than on the parts themselves and when a majority of the parts involved are social rather than technical in nature, we find that a focus on the numbers, on increasing the efficiency of parts without taking into account the function, processes, and structure of the whole is often counterproductive.

We learn that in order to increase the productivity of the whole, we frequently have to tolerate, even encourage what might be considered inefficiencies in the activities of parts so that they better complement each other. We learn that the carefully drawn lines between units, the precise job descriptions critical to successful analysis and numerical measurement, sometimes need to be blurred or even to be done away with completely.

It is not that measurement and the resultant efficiencies are not critical. They always will be. However, in order to produce the best results, this perspective needs to be absorbed into a much more encompassing whole. That whole is effectiveness which takes into account the non-quantitative as well as the quantitative aspects of improvement efforts; that considers the relatively unstable, unmeasurable human or social elements to be just as important as the measurable technical elements; and that tries to take advantage of that instability rather than ignoring or attempting to freeze it so that it does not mess up the equation.

In summation, the systems approach is a much more sophisticated starting point than the ones we have thus far built our efforts and models on in the quality improvement movement. At the same time, however, it is a much more logical one if we are willing to make the leap to do what is necessary to succeed in the long term.

But all we have been given thus far in our narration is history and theory, the historical roots and the conceptual foundations of the systems approach. Now it is time to move on to our current reality, to the nuts and bolts, to describe exactly how a systems approach-grounded quality improvement model is shaped, and how it must be implemented. That is the purpose of Chapter 3.

Topics for Discussion

1. How did your organization start its quality improvement effort?
2. If you have been through a downsizing, how did it affect your desire and ability to help improve productivity? If you have not, how do you suppose it would affect your desire and ability?
3. What qualifications and training should the head of a quality improvement effort have?
4. From a scientific point of view, why is systems theory considered radical?
5. Why do systems people define organizations in terms of function, structure, and processes?
6. What makes the concept of effectiveness richer than the concept of efficiency?

3 Five Phases Plus Two: Part I

After Reading This Chapter You Should Know

- The four characteristics necessary to a systemic quality improvement process.
- How to design the up-front familiarization phase.
- Why training in group process skills or statistical tools should not be part of or immediately follow familiarization.
- Why team building should immediately follow familiarization.
- What the five types of teams important to quality improvement are.
- Why a large number of quality team building efforts are illegal.

And Now the Hard Part

Let us start by looking back at our example of a typical quality process start-up to see if it was systemic in nature. First of all, was it a whole, a group of interdependent parts with one objective? Not really. We went through a familiarization phase, which included some up-front training. We got into team building after the training ended, but a lot was missing. For example, was *function* (long-range planning) part of the game? Was the effort tied into an organization overview? Did all the participants understand the role the organization wanted to play in the larger environment of which it was a part?

More important, were employees encouraged to contribute to the definition of the organization's function to give them ownership, to gain their commitment and to give them the necessary feel for how their jobs and ideas might fit into the whole? And finally, did a vehicle exist for effectively integrating the on-going contributions of employees into the long-range plan?

Again, not really. In our example if long-range objectives were even considered an important ingredient, they were delivered top-down; they were defined by management and presented as a given to the majority of employees. . . . So much for empowerment in that phase.

Next, we got into training, which falls into the *process* category. Experts were brought in to teach employees all sorts of things, but were these things what the employees needed? Did anyone bother to ask? Did hourly workers really need to learn about team building? After working in an office or on the shop floor for 15 years did secretaries really need to be taught how to interact with peers? At the age of 35 with a college degree and a family, did an accountant really not know how to solve problems?

Once again, the training was done to employees with no thought of seeking their input, their suggestions, and their ideas. The training was supposed to empower them, but to do what? The quality process was new to everyone. Employees had no experience with which to relate this training. So, nothing had actually changed culture-wise. Management was dictating what the training should include, at least indirectly, by bringing in consultants or in-house trainers to present canned packages. The empowerment was, at best, superficial; the level of commitment generated by it, at best, suspect.

Next came team building, which, from the systems perspective, provides the *structure.* In our example, was structure integrated with *function*? Were team efforts integrated with the organization's overall objectives? Well, if top-level management told participants what to work on, they were indeed. But, again, what about empowerment? What about finding ways to develop commitment? Is being allowed to work on a project, assigned by upper-level bosses, under the supervision of lower-level bosses really empowerment? Or is it just more of the same, perhaps dressed a little bit differently?

On the other hand, if top-level management did not assign projects, were team members familiar enough with their organization's overall objectives to effectively define their own? Did they have the necessary frame of reference into which to fit their efforts, one they understood and accepted because they had helped shape it?

Next, was team building integrated with training so that the relationship between the two could be more easily defined by those directly involved? Once again, the answer has to be no. By the time team building began the training had long since been completed, maybe months or maybe years earlier. Most of what employees had learned had been forgotten. Team builders were left, pretty much, on their own.

And finally, concerning team building, was a comprehensive network of teams put into place? If one existed, was it well-integrated, did each team know what the others were doing so that repetition did not occur and so that the teams could contribute to each other efforts? Was the team network a system or was it simply an aggregate of disconnected pieces, each doing its own thing, linked, if at all, only by those at the top of the organization?

Function, process, structure—from a system's perspective they obviously did not exist in our example. They do not, in fact, exist as defined in a majority of quality improvement efforts. Only pieces exist, and the pieces are rarely integrated to the necessary extent. Neither do we find efficiency or effectiveness in our example. In terms of efficiency, our defensiveness has kept us from discovering the one best way to organize quality improvement efforts. In terms of effectiveness, this same defensiveness has kept us from taking full advantage of the value-rich input of those upon whom success is most dependent, the employees. We have kept our quality improvement efforts *top-down* and allowed only *semi-empowerment,* when the key to achieving the desired results in such efforts is *all over at once* and *full empowerment.*

What It Takes

In order to be successful, from a system's perspective, a quality effort must have several characteristics. First, it must be participative, really participative. This is the only way to truly empower employees and, thus, to gain their commitment. This is the only way to truly integrate the effort and to make it effective as well as efficient. (Actually, as we know effectiveness incorporates efficiency. It just factors values into the equation as well as numbers.) Employees must help shape every phase. They must help create the effort, rather than simply being handed management-defined assignments. They must help formulate long-range objectives. They must help design the involved training. They must control the activities of their teams.

Second, a quality effort must eventually be organization-wide and well-integrated. It must tie function, processes, and structure together. The quality effort must itself be a system. It must create a system of quality improvement teams and task forces.

Third, a systemic quality effort must be ongoing. And if the whole of the effort is ongoing, the parts must also be ongoing. The planning phase, for example, must not be a one-shot deal. The involved processes, including training, must continue as the organization evolves. Teams in the network (structure) cannot pop up, work on something, then disappear. They must be stable and help tie the effort together.

Finally, a quality improvement process must be capable of learning constantly and improving itself. It cannot do this in the most effective way unless it is participative, organization-wide and integrated, ongoing. It cannot do this unless function, processes, and structure have been melded into one close-knit, interdependent core.

On to the Model

But enough foreplay. It is time now to describe just how a systemic quality improvement process can be put together, what the pieces are, and how they must fit, taking all the requirements we have laid out into account.

All systemic quality improvement efforts must include five integrated phases—familiarization, long-range planning, team building, training in individual and group process soft skills, and training in the use of statistical tools when appropriate. All must also include two appendages—development of an appropriate means of measuring the success of the quality effort and development of a reward system that helps strengthen commitment to it.

So how do we implement organization change from a system's perspective? We should have realized by this time that while the underlying concept—*the whole is more than the sum of its parts*—is seemingly easy to understand, it is not going to be at all easy to introduce it to organization cultures raised on generations of analysis, fragmentation, and the resulting individualism and in-house conflict.

In real life, then, we usually need two things to succeed. The first is a crisis. In terms of this requirement, it is becoming increasingly obvious that in the United States we are moving right along. Despite disclaimers by those benefiting from the current situation, all we need to do is look around the workplace, around our homes, around our communities to see the negative effects of our current practices. The symptoms are all over.

The second thing is the right vehicle. We also have access to that. It is the quality improvement movement, but we have to shape it correctly. We have to pull all the pieces floating around out there together into a meaningful, systemic whole that relates them directly to function, structure, and process. Long-range planning obviously fits with function or the definition of organization purpose. Team building has to do with structure, with what the quality team network should look like, how quality teams should be integrated with each other and with task forces created on the operational side. It has to do with whether or not a labor-management council should be formed and how it should fit in. It has to do with what the facilitator network should look like.

Familiarization, group process skills training, and the introduction of measurement techniques useful in the workplace all come under the category of process and have to do with maintaining flow toward realization of the desired organization characteristics and purpose.

Just like function, structure, and process, these five pieces of the quality puzzle are totally interdependent and cannot be separated without destroying the critical, unique characteristics of the whole. They have to be thought of and dealt with as a system. None of them can stand alone and succeed. Familiarization sets the tone. Team building provides the vehicle through which employee expertise and creativity can be unleashed most effectively. Without proper training, however, team efforts will come to naught. The long-range planning phase provides the necessary frame of reference into which everything must fit. Yet, long-range planning is affected by team activities. These activities also help dictate the types of training necessary. They also affect the tone of ongoing familiarization.

While the pieces, themselves, of a systemic approach do not differ greatly from those used in most quality improvement efforts today, the interactions that occur within and between these pieces do. The differences might be subtle in some cases, but they are critical. First, however, let us make sure that we understand the five pieces.

Talking It Up

The familiarization phase of a systematic quality improvement processes (QIPs) must always come first. During this phase upper-level management, quality department staff members, and consultants explain to the entire work force the need for improved quality. They usually begin this process at the top of the organization, working their way down gradually to the hourly ranks. They discuss the fact that after World War II, when the United States had little competition, quality was not a major concern, but now that European and Asian nations have caught up technologically and are cutting into our share of the world market, quality has become the decisive battleground.

They help employees realize that a successful quality improvement process is just that. Rather than a project or an exercise with an end, it is an ongoing process that eventually causes the corporate culture to change, to become more participative, and better integrated.

They explain the difference between quality control, which basically checks for product defects and then tries to discover and eliminate the causes of these defects, and the new, more comprehensive quality thrust that realizes that better product quality results from improvements in the quality of manufacturing processes, management systems, and the work environment as well.

They stress that the ultimate objective of QIPs *must* be to improve the bottom line, that if the corporation fails financially, the rest is meaningless. They point out that while our traditional emphasis on producing increased quantities, frequently at the expense of quality, does not always lead to such an improvement, a continual emphasis on quality that, in the short run, might even necessitate a cutback in production levels, does.

They discuss the fact that improved quality results primarily from a high level of employee commitment and that three things are fundamental to this high level: management emphasis on insuring job security, reasonable levels of pay, and respect for employee work-related ideas and needs.

They discuss the fact that upper-level management cannot dominate or channel the process, at least initially, but must show strong support for it, not only with words but with actions. They discuss the fact that senior executives must function as a resource to employee improvement efforts; that upper-level management must constantly encourage and applaud these efforts; and that, most importantly, the CEO must set the example by more fully utilizing the talents of his or her *own* direct reports in the formulation of corporate and department policy.

They talk about the need to involve unions in such efforts, with local representatives being encouraged to play an active role as facilitators, resources, and cheerleaders. A properly run QIP can do nothing but improve the situation of employees. Progressive union leaders understand this and actively support such efforts. Those who resist do so for three reasons:

1. They have experienced such poor relations with management that they oppose any sort of collaboration. They are afraid the effort is a ploy and that, once the union's power is broken, management will revert to its old ways and take unfair advantage of employees.
2. They do not believe that employees are capable of representing themselves adequately in labor-management negotiations.

3. They are afraid of losing their leadership positions as the process draws labor and management closer together.

QIPs have many of the same objectives as unions, although a QIP's approach may be different. By different we mean that while many union leaders—especially in the United States—continue to think in win-lose terms as do many corporate executives, QIPs advocate a win-win approach. Also, QIPs encourage employees to think and speak for themselves. The union leader's role shifts from representing membership directly to supporting, encouraging, and safeguarding member's efforts to voice opinions, make contributions, and satisfy needs.

They explain the team building model that has been picked. They stress the fact that the team building process will be action rather than training driven. Teams will be formed immediately and will begin working on projects within a week. What employees, especially lower-level ones, need most is a chance to perform, someone to keep them on track, a positive response to their requests, and a rapid initial taste of success to convince them that they, indeed, have been given the power to create change. Up-front training in problem identification, problem-solving, and group process techniques is usually a waste of time. Employees already know the problems because they live with them. Also, they often have a pretty good idea of excellent solutions or, with proper support, are quick to ferret them out.

At the same time, familiarizers try to make clear that what is about to be attempted will take a long time; while positive changes will occur almost at once, it might take years before the sought after new culture is finally in place.

They end by making the point that the process will survive only if it proves beneficial to employees as well as to the company. At this point, the company asks only that people give it a chance. If it does not work, the company has given participants the power to change it, or to shut it down. A fairly safe bet would be that every single employee can think of at least one way to improve a product, manufacturing process, management system, or the work environment. The QIP will provide the opportunity to turn these ideas into reality.

In summation, the familiarization phase is self-defining. Its purpose is to offer the *why* behind a QIP and to explain the *what* we need to put into place to accomplish our objectives. It is ongoing. Presentations are made continually, at all employee levels, with shifting emphasis depending on the questions that arise from participants and on the quality process stage. Tools used during the familiarization phase can include posters, videos, presentations, stickers, visits to other sites—anything that will help employees focus on improved quality as a primary workplace objective; that will help employees better comprehend the power being given them and better understand the organization's customized approach.

Too Much Too Soon

Once the familiarization phase is well under way, most corporations, from a systems perspective, get side-tracked. Most commonly they jump, as we have said, from familiarization directly to group process training. One of the major reasons QIPs lose themselves so quickly in group process training is that most quality consultants and members of QIP departments attempting to deal with the socio, or human side of the equation, vs. the technical side have backgrounds in organization behavior, organization development, or psychology. Their skills lie primarily in working directly with individual's problems and group processes. It is natural, therefore, for them to take this approach.

Up-front training, however, can cause serious process-related weaknesses. For one thing, such training, in large companies at least, is usually top-down. Professional trainers, following the lead of those who organized the familiarization phase, begin by training managers, who, in turn, are supposed to train everyone else with the support of the professionals. There is, however, a critical difference between the familiarization phase and the training phase. While the up-front portion of the former can be presented to large audiences and completed in several weeks; the latter, when dealing with a company of any size, ultimately involves running several thousand students through well-organized, two- to three-day sessions. Such an effort is extremely drawn out so that by the time that everyone is trained, many of the earlier students have lost their enthusiasm, their workshop notebooks, or both.

At the same time, such training is rarely, if ever, adequate. Learning a technique in the classroom, even practicing it there, never gives students all the answers or prepares them fully for the real-life situation. A tremendous amount of support, therefore, is necessary when those initially trained begin passing down their new knowledge and skills to lower levels. Such support, however, is rarely available. The corporate quality staff and consultants can visit just so many work sites during the year and can answer just so many phone calls.

Also, many middle managers see QIPs as an attempt to get rid of their jobs. They have heard about too many layers of management and about the push to get hourly workers to make more work-related decisions so that some of these layers can be eliminated. Giving middle-level managers responsibility for passing on problem-solving and decision-making skills is, in many cases, like asking the last werewolf to teach the pretty lady how to shoot silver bullets. It does not make sense. Inexperienced or emotional supervisors create a fuss or refuse openly to cooperate. The old-timers go through the motions, but then make sure that nothing happens.

Finally, when a comprehensive training effort is mounted before the team vehicle is put into place, employees, as we have said, receive little or no chance to contribute to the design of the package. What they get, therefore, is frequently not what they need. As a result, the employees must either go without as a result of their reticence to complain when management is so enthusiastic or they must be trained again at some later date.

The second detour companies usually make following the familiarization phase is that they jump directly into the introduction of statistical measurement techniques (SQC and SPC). Again, the belief is that the involved education effort will help generate employee commitment. It will not. What many of us have done is to misinterpret the producers of W. Edwards Deming's success in Japan. One of the major differences between the traditional Japanese and U.S. work forces is that with the former, commitment to the good of the organization as a whole is a given. It is a cultural characteristic. The problem that Deming faced, therefore, was not to *foster* but rather to effectively *channel* an already existing sense of commitment and team work. In this situation, SQC was a useful tool in the right place at the right time.

Owing to the traditional adversarial relationship between labor and management in the U.S. work force, however, we must begin at the beginning. The first step must be to generate the necessary level of commitment to improved quality. Without it, the introduction of measurement techniques will be seen, in its best light, as having nothing to do with the betterment of the worker's situation. At its worst, it will be seen as a threat to job security and something that must be blocked.

Bring on the Teams

The team building phase should follow familiarization. The key is to begin producing positive change as quickly as possible. Let us say that a company is one of the small but growing number which understand this and lines the quality process pieces up right. Let us say that it is willing to seriously try to empower employees through team activities. Despite this giant step forward, chances are that its quality improvement effort will still fall short. The effort will fall short because its team building effort, although headed in the right direction, is not designed correctly.

In this section we shall focus on getting the details of the most important phase of any quality process right. This phase is team building. Team building efforts must take the lead. While the other phases of the quality whole are critical, the team network is more so for the following reasons:

1. The teams provide a visible structure for the entire process. They form a focused framework that supports the ongoing familiarization, training, measurement technique introduction, and strategic planning phases.
2. QIPs ultimately live or die, as we have said, by the bottom-line improvements that they produce. If the teams in the network are properly organized, each will begin working on and completing projects two to four weeks after it is formed. The process will soon produce measurable financial returns, thus relieving upper-level management's anxiety about the bottom-line value of the exercise.
3. Again, commitment from employees on all levels is the most necessary ingredient to a successful QIP, and team activities, more than anything else, help generate that commitment.

Despite these realizations, it is during the team building phase that most processes eventually begin to flounder. One reason for failure is management's aforementioned reticence to let go, to truly empower team members. Managers struggle, sometimes openly, sometimes covertly, to maintain control. Their fear, frequently justified in shortsighted organizations, is that if the team building effort succeeds, they will become expendable.

There are, however, other, more technical reasons for the failure of team building efforts that have less to do with the workplace culture and personalities and more to do with our learning curve. We are relatively new at this systems, quality improvement, employee empowerment stuff. Despite all our frustrations we are doing pretty well. We have learned quite a bit in a relatively short period of time, but we still have a way to go.

One thing we have to learn is that at least five different types of teams, committees, and councils exist which affect and should eventually be part of successful quality improvement efforts. Each has a specific, well-defined role. None of the five can be mixed. Yet, we continually try to do so, not intentionally, but because we do not yet understand the boundaries. And in mixing these teams we not only scuttle the process, but, in many cases, actually break the law. In fact, if we were to survey the field my guess is that we would discover most current quality team building efforts to be illegal.

How's that for a surprise?

The Five Musketeers

Let us start by identifying the five types of employee involvement groupings and each one's individual purpose. The team that appears most frequently in organization improvement efforts is the traditional *management directed, project specific task force.* A manager, usually upper level, decides to build a team around a specific project. The manager defines the skills required to complete the project, picks a leader, usually a lower level manager, then brings together the proper mix of employees. The task force completes its assignment, then sends its recommendations to the person who built the team. Based on these recommendations that person decides what should be done. The task force is then disbanded or given another project which fits its configuration.

The strengths of this type of team include its direct tie-in with corporate department objectives, the ease of team formation, and the degree of control exercised by management. Weaknesses include the fact that its purpose is to advise and make recommendations. It is not usually involved in defining the problem to be addressed, in reaching final decisions, and, frequently, not as a unit, in implementation. Participants, therefore, do not gain a true sense of project ownership.

The second type of team to materialize historically, which we introduced in Chapter 1, was the *workers council,* now often called a *labor-management committee.* This type originated in Eastern Europe and showed up in at least five Western European countries—Norway, France, Sweden, Holland, and Germany—between the end of World War II and 1950. The workers council is a group of laborers and managers that meet periodically, usually monthly, to discuss both technical issues and working conditions. It is found in unionized organizations. The representatives of labor are union officials and, therefore, present the union viewpoint.

Workers councils were initially formed as a means of improving communication between upper-level management and the workforce as the size of organizations and the number of managerial levels grew. They also gave labor the opportunity to develop a more holistic perspective of the operation. When the involved unions were adequately represented, the councils provided a means of dealing with employee/management complaints informally as a way of avoiding the filing of grievances and costly arbitration. Finally, unlike management directed, project specific task forces, councils defined their own agenda, both sides bringing issues to the table.

The weakness of the workers council concept is again that participants can only make recommendations. They are not involved in the final decisions. When, however, team members reach consensus on a modification that will benefit everyone, those who do make the final decision usually listen.

The third type of team to appear historically, which was also previously discussed in Chapter 1, was the *autonomous work group,* now sometimes called a *self-directed work group.* It originated in the coal mines of England during the 1950s as a means of making work more interesting and challenging and of increasing the level of employee commitment to improved productivity. The concept then spread throughout Europe, the most publicized effort being that mounted in Sweden. For the first time, team members did, indeed, make the final decisions and control implementation.

Autonomous work groups are self-directed, manager-less teams of employees with matching skills working together on a daily basis to generate or exceed a set quota of finished parts or services. They decide what to do. They decide who should handle each task daily. When cross-training is appropriate, they are responsible for delivering it. They take the lead in solving production-related problems. They pick replacements if members leave the team. They decide if and when members *should* leave the team. With autonomous work groups, managers, for the first time, become facilitators and resources rather than watchdogs and decision makers.

The strength of the autonomous work group is that it gives control to the people who know the job the best. The weakness, if any, is that team members focus on their own activities and sometimes lack a sense of integration with the rest of the operation.

The fourth type of team deals with this weakness. It is a product of the system's school of thought. Its major architects are Russell Ackoff and Jamshid Gharajedaghi. Ackoff calls it the *circular organization.* Its basic component, a system of interactive boards, has been put into place at Anheuser-Busch, Alcoa's Tennessee operation, Kodak, Metropolitan Life, the Super Fresh grocery store chain, Armco Latin American, and the Ministry of Public Works in Mexico.

The major purpose of the circular organization, according to Ackoff in Chapter 4 of *The Democratic Corporation,* is to:

1. Encourage the circularity of power, or true industrial democracy.
2. Give each employee the ability to contribute to decisions that affect him or her.
3. Give board members the ability to directly make decisions that affect no other stakeholders.

Each manager at every level in a circular organization has a board. On it sit all the employees who report to that manager, as well as the manager's supervisor, and any other stakeholders, in-house or outside, considered necessary to the issues addressed. Members of this latter grouping may attend on a permanent or on a temporary basis.

Each board is responsible for its unit's planning and policy-making. Each is responsible for coordinating plans and policies on the same level, for integrating plans and policies with both lower and higher units, for dealing with quality of work life issues raised by board members, and for evaluating and helping enhance the manager's performance. The boards usually meet monthly and, obviously, end up running the company.

This arrangement, coupled with the autonomous work group, approaches the ideal in terms of participative management. Most organizations, however, are not immediately capable of accepting such a radical change. They need a *bridge,* a middle stage that allows them to make the transition more gradually, to acclimate. This bridge can be the *quality improvement team,* the fifth and last type of team important to successful quality improvement processes.

The quality improvement team allows hourly and professional employees, as well as managers in many cases, to grow accustomed to empowerment by working first on improvements in their own areas of expertise, then expanding slowly outward until they can eventually become active in the formulation of plans and policy, in actually helping to run the company. At the same time, this bridge arrangement gives managers the chance to learn that employees are capable of making contributions and decisions before involving them in the critical operational ones. It also allows managers to work out their new roles as facilitators, without immediately coming under the gun as the focus of one of Ackoff's boards.

The quality team network serves as a sort of proving grounds, but one that almost immediately begins producing desirable results. Representatives from all different functions—accounting, maintenance, production, secretarial, and so on, should belong to teams in a well-integrated network. We are not talking about cross-functional teams. Task forces are cross-functional; quality improvement teams are not.

The primary purpose of quality improvement teams is to develop commitment. This is done by allowing them to work on the issues most important to members. Such issues are invariably the ones which affect the members most directly and continually, the ones found within their own 25 square feet. After these issues are resolved, boundary issues can be addressed.

When quality improvement teams are made cross-functional, they have two choices, both undesirable. The first is to begin working on broader issues immediately so that everyone is involved, skipping over those most important to individual functions. The second choice is to work first on an issue affecting only one function, then on an issue affecting a second function, so that team members end up spending most of their time solving other peoples' problems instead of their own.

Teams should start out representing one function, or, if the function is large enough, part of a function, so that members can get their own sandbox in order before worrying about anyone elses, before expanding their horizons. Once individual functions are running smoothly, boundary issues can become the priority, and teams can combine temporarily or form task forces to deal with them.

Teams must be formed in all parts and on all levels of the organization. In order to generate the desired degree of ownership and commitment, at least initially, hourly workers, professionals and managers should have separate teams. This arrangement will keep team sessions from turning into business as usual, with management taking the lead and defining the priorities.

The systemic quality team network has three integrated levels. On the first are the hourly/professional teams defined by function. On the second are the managerial teams defined either by function (including all managerial levels from that function except the top-most) or by level (including all managers from across the organization on the same level). On the third level of the quality team hierarchy we find the lead team which included the CEO or facility manager, his or her direct reports, and the head of the quality team facilitator network.

Quality network teams should meet according to a set schedule. They must be led through an initial exercise to help them identify and prioritize the problems on which they want to work. Their members should communicate team activities to those they represent and get their input. Hourly teams must be protected from managers who remain opposed to and attempt to frustrate their efforts. Team activities must be fully integrated on all levels. The team effort must have a set of ground rules that are well-thoughtout, understood, and agreed to by everyone.

Finally, team building efforts should start at the shipping dock or the sales counter and move from there backward through the product development steps. This is important for several reasons. First, if this approach is used, customers are more likely to benefit from early improvements. Second, most of the problems identified by teams concern their relationship with the feeding rather than the following function. For example, the packaging unit is certainly more concerned with what it receives from the manufacturing unit than with what it sends on to shipping. The idea is to have the team building phase flow in the same direction as team interests.

The strength of the quality team network is that it eventually involves the entire workforce in upgrading the operation in a well-integrated manner. It also provides a tremendous stimulus for intra-organization communication and for cross-education. The weaknesses are that it is designed solely to generate improvements, not to run the operation like autonomous work groups and circular organizations, and, as we have said, that most such networks are not built correctly.

Five Legs of the Stool

The five types of teams we have defined serve in five different organization dimensions:

1. The management directed, project specific task force helps managers do their jobs and meet their objectives in the most efficient manner.
2. The labor-management committee helps bridge the communication gap between upper-level management and the workers and between management and unions.
3. The teams in a quality network, both worker and managerial, involve all employees in making improvements in every facet of an organization.
4. The autonomous work group, if properly supported, helps workers do their jobs and meet their objectives in the most effective manner.
5. The circular organization allows companies to be run democratically and to make the most complete use of employee expertise on all levels.

The first three types of groupings complement each other and make it possible for the fourth and fifth group to evolve. There is no reason that these three cannot exist simultaneously in an organization. At the same time, however, there are several good reasons why, as we have said, none of these three can or should be combined. The problem comes when the characteristics of the task force, the labor-management committee, and the quality improvement team are confused. This is also where we get into legal trouble.

We are all familiar with the National Labor Relations Act (NLRA). It was passed to help balance bargaining power. It was passed to protect employee rights and to ensure that they receive a reasonable share of this power. It was passed to protect them from, amongst other things, being taken advantage of through the creation of sham unions.

Management, in a non-union company, has been known to tell workers to form in-house groups, usually of volunteers, to represent the workforce and to formulate recommendation on issues concerning conditions of employment. Management obviously dominates such processes and may have no intention of honoring employee recommendations. If this is true, the groups, lacking the legal and strike-related leverage of unions, have no recourse. They must play the game.

However, because such in-house groups 1) Represent the workforce, 2) deal with what can be considered negotiable issues, 3) are management dominated, they can be classified as labor organizations or sham unions by the National Labor Relations Board (NLRB) and ruled illegal according to the NLRA. It is important to understand, however, that the way these three factors are combined is what makes them illegal, rather than the individual factors themselves.

Based on what we have just said, it appears that only one of our three types of teams can comfortably address conditions of employment or negotiable issues. That is the labor-management committee in a unionized operation. Such committees can represent the workforce and cannot be dominated by management. If their recommendations are rejected and the reasons given are not acceptable, union members can always revert to the traditional path and file a complaint.

Most of the confusion we are talking about involves the interface between management directed, project specific task forces and quality network teams. The former are obviously dominated by management and, therefore, are immediately suspect when assigned negotiable issues to work on. The 1992 NLRB decision on Electromation Inc. is a case in point. Management of this non-union operation formed volunteer action committees of hourly employees to generate recommendations on specific management prescribed topics including absenteeism and infractions, pay progressions for premium positions, and attendance reward programs. A manager sat on these committees and contributed. At the same time, a union was trying to organize the site. It filed a petition for an election, which it lost. Afterward, it filed an objection, claiming that the action committees were illegal labor organizations. Electromation Inc. became a test case, which the union won (Schlossberg and Reinhart, 1992).

Management should not try to dominate quality network teams and tell them what to work on. When it does, it crosses them with management directed, project specific task forces and jeopardizes the process by taking team ownership away from members.

Quality teams composed solely of hourly or professional members do frequently pick projects involving working conditions. Shop floor hourly teams in primary industry, for example, normally put improvements concerning safety at the top of their lists. This is legal because there are no managers on the teams. Also, hourly teams usually deal with specific instances—a loose railing, a faulty connection—rather than overall policy. Policy issues are left to management teams.

If the operation is unionized, one of the previously mentioned quality team ground rules requires that all who will be affected must agree to recommended changes before they are made. This includes the unions. The easiest way to deal with such issues might simply be to form a labor-management committee to support task force and quality process team efforts. Questionable issues and projects raised by either could be channeled to it.

When we first read about the NLRA and the rulings made by the NLRB we were upset. We thought the Board was trying to block the quality improvement movement as a threat to the union movement. Now we realize that we were wrong, that the NLRA can help push us in the right direction. We doubt whether this was the rule's original intent, but that does not matter. The wording of the rule can be used to stop upper-level management from dominating team activities. It can be used to stop upper-level management, to stop management on any level from dictating the projects teams work on. It can be used to help make clear the need to fully empower hourly employees, to give them their own teams where they can define their own projects.

In essence, knowingly or not, the act helps push us toward the facilitator style of management. It also gives unions the opportunity to play a key role in changing the nature of the workplace. It gives them the power to help make sure quality improvement efforts are being built in the right way.

Topics for Discussion

1. How many of the four characteristics necessary to systemic quality improvement does your organization's effort have?
2. Why does the familiarization phase in the systems model not go into more detail?
3. What are the reasons that up-front training is usually ineffective?
4. What key differences exist between the five types of teams discussed?
5. How can the five types of teams be effectively integrated into a long-term quality improvement effort?
6. Why does the NLRB shut down some quality improvement efforts?

4 More on Team Building

After Reading This Chapter You Should Know

- The major reasons most team building efforts fail.
- The difference between ground rules and rules of team etiquette.
- How to truly empower employees without losing control.
- How to protect team members and projects.
- Three problem-solving techniques useful to task forces.
- Two necessary start-up techniques for building quality teams.

Ground Rules—Quality Improvement Process Glue

During our discussion of teams we have spoken continually of *ground rules* critical to their success. While ground rules were designed initially for quality improvement teams, at least some of them have since been adopted by task forces and labor-management committees. And as they seep down from the domain of these three types of teams into the general workplace culture, which they inevitably do, as employees begin applying them to their normal workday activities and responsibilities, these ground rules have helped prepare work forces for the successful introduction of autonomous work groups and of the circular organization concept, of the hierarchy of boards on the everyday operations side (as opposed to the meeting one hour a week to make improvements side).

Ground rules, if they are to be truly effective, must eventually be absorbed into the corporate psyche. They cannot be learned in a classroom-type setting. Team members, therefore, are not expected to fully understand the importance of ground rules or even their meaning when introduced to them during the initial familiarization sessions. The bulk of learning concerning ground rules is done when teams begin struggling with actual problems and improvements. Team members grow into them, sometimes even reinventing them as the need for each arises.

In most cases, the ground rules we are talking about are not used. This is a major reason why team building efforts are not producing the desired results. In such instances, the lack of success has been attributed to the following:

- Team members miss meetings owing to production crises and other demands on their time and have difficulty getting involved.
- Management sees the team network as a vehicle for accomplishing their own objectives and begins dictating team projects or taking over team meetings.
- Teams eventually bury themselves, becoming involved in so many projects at once that nothing gets finished and members lose interest.
- Teams take on projects that are too difficult, too slow moving, and too long term so that members again lose interest.
- Team meetings degenerate rapidly into gripe sessions.
- Team members start squabbling, or factions develop with conflicting interests.
- Teams function for several weeks or months, then disband, as members become disillusioned by the lack of support for or the negative reaction to their ideas.
- Teams get in trouble with other teams and departments because the changes they make have unexpected, negative consequences on other parts of the operation.
- Actual and suggested changes generated by teams are seen as a threat by supervisory personnel and are blocked.

The missing piece that can prevent all of the above from happening is a well-thoughtout, well-defined set of ground rules agreed to by everyone from the CEO or facility manager on down. There must also be a mechanism built into the ground rules that ensures that they will be obeyed by everyone who is affected by or who is capable of affecting the process. Such ground rules are the glue that holds the team building phase of a QIP together. However, because the team network can be used as the vehicle to continue the familiarization phase, for both definition and implementation of necessary technical and management skills training, for the introduction of measurement techniques, and for the strategic planning effort, the glue that holds the team network together is actually foundational to the entire comprehensive effort and must contain the right mixture of ingredients.

Most quality process team building efforts do, indeed, have a set of rules by which they operate. These include such things as "No interrupting." "Putting down someone else's idea is unacceptable." "Team members are not allowed to leave to answer phone calls." Such rules, however, are not ground rules. Instead, we would call them *rules of etiquette.* They focus solely on the interaction between participants. *Ground rules are a set of rules developed by systems approach practitioners that tie team efforts into the rest of the quality process and into the organization culture as a whole. Few of them relate directly to team member interaction. Rather, they have to do with the interaction between the team as a whole and the larger environment of which it is a part.*

In Chapter 4, therefore, we shall introduce those ground rules most commonly used in systemic quality improvement efforts. We first came across them in literature describing the contribution Eric Trist and his staff made to the Jamestown, NY project, perhaps the first in this country to introduce the concept of quality improvement on a community-wide basis. Systems practitioners have since expanded on and used them successfully in a wide range of organization change efforts including those started in primary industry, banks, hospitals, government organizations, and schools. In every instance, participants were eventually allowed to modify the ground rules. In every instance they ended up with basically the same set. With this in mind, what we might call the *core set* of ground rules will now be presented.

Ground Rules

1. Teams must meet, at least initially, on a regular basis, and members must be allowed to attend.

As proof of its commitment, management must allow teams to meet during working hours. Forcing employees to add time to their working day for meetings, even if they are paid for that time, shows blatant disrespect. Commitment will decrease, rather than increase.

When a during working hours team building phase begins, many managers are reluctant to release employees for the 60 minutes per week required. Work schedules are disrupted, responsibilities have to be reassigned. Quality team meetings are usually a low priority. We have never worked with an organization where at least some managers did not say that it was impossible to release staff. Telling them that the process will produce improvements is not sufficient. They have to be shown. Several meetings will pass before the teams start generating notable results. Therefore, managers frequently must initially be forced to arrange coverage for process participants unless a true crisis occurs.

In some cases, the manager does not have to make the arrangements. The employees in the unit, if given a chance, find ways to cover for each other. We also have never worked in an organization that did not find a way to make employees available once the team process had proven its value.

After six months or so, when the team network is established and teams have completed a number of projects, members can themselves begin deciding how frequently they need to meet. Sometimes they cut the meeting back to one-half hour. Eventually, the teams begin meeting every other week, unless something important comes up. Finally, teams meet only when they feel the need.

2. Original team members must attend at least the first five meetings.

This ground rule ensures that team members get a chance to learn the value of the process before deciding whether or not to remain involved. When the first teams are brought up, asking for volunteers frequently does not work. No matter how thorough the initial familiarization session has been, hourly workers/professionals see the meetings mainly as an additional responsibility. Potential benefits of the process are suspect. Most employees have long since given up hope of seriously impacting the operation with their ideas. Also, you frequently get the complainers.

Managers should be asked to pick the employees they think can make a real contribution. Mandatory attendance at the first five or six meetings gives team members a chance to understand what the teams can accomplish. By that time two or three projects will be underway, if not completed. At that point, if members still consider the meeting a waste of time, they can ask to be replaced.

Some people, as we all know, never do buy in. They should not be allowed to interfere with the process, but, at the same time, they should not be forced to participate. The quality improvement effort must prove itself to employees. Each individual must see the benefit to himself or herself. This is the only way to gain true commitment. Once the process begins producing positive results most doubters will become active, but forcing them to do so sends the wrong message.

3. Teams must find substitutes for members who are on vacation, on sick leave, or who are absent for any other reason.

Another way to involve more employees is to require team members with legitimate reasons for missing a meeting to send a substitute. This requirement also encourages team members to keep the co-workers that they represent informed of team activities. If they do not, it will be much more difficult to find a substitute when needed. In some instances, substitutes become so involved that the team designates them as regular alternates. In others, it is necessary to expand the team to include employees who want very much to participate on a full-time basis and who have proven their worth as substitutes.

4. All team minutes are confidential unless released by the team.

Paperwork is kept to a minimum in a systemic quality improvement process. In fact, the only two official pieces generated are the team meeting schedule and the minutes of meetings that list, prioritize, and track the progress of all projects.

Another means of giving teams an enhanced sense of ownership is to allow members to decide if, when, and to whom they wish to release their minutes. Initially, team members are reticent to do so for fear of retribution. They do not want people to know what they are working on so that at the first hint of danger they can shut it down without too much damage being done. Forcing teams to share their minutes, therefore, especially at an early stage might arouse suspicion as to upper-level management's motives. It might dampen enthusiasm, breed caution concerning project choice, and negatively affect the process.

Once participants start identifying project stakeholders and getting input, however, they realize quickly the need for and the value of allowing as many people as possible to know what they are working on. Therefore, after several weeks, teams are usually eager to disseminate session notes, especially to upper-level managers, as a way of opening a channel of communication to them.

One copy of all team minutes is kept in a central file maintained by the head facilitator. At the same time, lists of the projects each team is working on and has completed are displayed in a central location on poster-sized sheets of paper. These lists are accessible to all so that the entire work force knows what is going on, so that everyone can watch the number of projects completed grow, so that anyone who thinks he or she will be affected by a project but has not been contacted yet as a stakeholder can speak up.

5. Teams must be allowed to identify the projects they want to work on themselves. They must also take responsibility for implementing the changes proposed.

Probably the least popular feature of the systems approach to quality improvement is its refusal to combine hourly/professional employees and managers on teams. The second least popular feature is its insistence that teams be allowed to identify their own projects. The belief voiced is that the major purpose of teams is to improve the relationship between levels by allowing representatives to work together on projects in a neutral arena, and that upper-level management, owing to its overview, knows best what these projects should be.

This type of thinking misses the point. The most important objective of successful quality improvement efforts, as we have already said numerous times and will repeat numerous times more, must be to generate employee commitment. This is done by management showing respect, by management trusting hourly workers/professionals enough to give them their own teams where they can come up with their own ideas and projects.

When teams are given the right to identify their own projects, some supervisors fear that time will be wasted on relatively superficial improvements, that members will focus on creature comforts. If the team is started correctly, we have never seen this happen. Another fear is that the projects they pick will have nothing to do with the reality of the organization. We have never seen this happen either. It is rare that by the time the network is completed some team has not picked most of the projects managers would have assigned and has not identified several others, equally as important, that management has not thought of. And, as a fallback position, if a project a manager considers important is not being dealt with, he or she is free to form a task force around it at any time.

Another value of allowing teams to identify the projects that they wish to work on concerns ownership. Most corporate heads agree that employees work harder on improvements for which the employees themselves have defined the need, and that they are more eager to see such projects through to a satisfactory conclusion. The issue of ownership is also why teams must be made responsible for taking the lead in implementing the improvements designed. They must carry projects all the way through, rather than coming up with a solution or alternative solutions (as task forces do), then turning it or them over to management for a decision and implementation. When a large number of teams are involved, this latter course of action would put managers in an impossible situation. They would not be able to deal with the number of proposals pouring in and the process would, at least, stall.

6. Teams must prioritize their problem list and work on correcting the problems one at a time until an acceptable stage of progress is reached on each. They should be encouraged to start with the simplest.

One major concern of the team building phase is keeping members focused. They jump from one problem to another, defining a few action steps for each, but completing very little, as new projects are continuously introduced and pieces of the old ones are lost. Once members have compiled the initial list of problems and improvements that they want to address, they should be instructed to pick one project and work on it until all action steps possible at that point have been defined and assigned. Only then can a second project be chosen and work started. Working on more than two projects at a time, however, is strongly discouraged.

Especially when the team starts up, members assigned action steps to complete during working hours forget them, or do not get around to them. This let-down must be addressed immediately by the facilitator. Team members have been empowered, but that empowerment includes responsibility. This is their chance to make a difference. It is up to them.

Facilitators should encourage members to begin with relatively easy projects. The objective is to produce quick results so team members realize that they can indeed have an impact and so that supervisors realize the teams are capable of effectively addressing lower-level, day-to-day, nuts and bolts issues, thus releasing the supervisors to focus on the larger, more systemic challenges that they had been wanting to address for months or even years.

7. Team projects cannot be taken over.

Another surefire way to kill a team building effort is for managers to "jump the process." A team identifies a project and starts working on a solution. The team needs input and invites a manager or expert to attend a meeting. The manager or expert becomes interested in what the team is working on and decides that his or her department can deal with the issue more quickly and effectively. The manager, frequently with the support of superiors, tells the team, "We are taking charge now. Thank you for your input and efforts. Work on something else."

This is a short-sighted strategy. It destroys the sense of process ownership and, thus, the commitment of team members; but, perhaps more important, it sacrifices long-term savings for a quicker short-term turn around time. It is true that in many cases the manager can probably complete the initial project more rapidly. But that is just one project. If the team building process is properly nurtured, in a relatively short period of time the organization will have 30 to 50 teams, each completing several projects a month. The numbers are obvious.

Absolutely no one is allowed to jump the process. If a project affects other functions, ground rule number 12 requires that they be involved; but they are not allowed to take the project over. The team that addresses an issue first takes the lead. It can turn that project over to another team or a task force if it wishes, but that decision must be made by the originating team.

8. All team members are equal. All decisions are made by consensus.

Some quality improvement teams make decisions by vote. While time might be saved, the result is usually not the best. Those against the chosen solution have, or at least believe that they have, good reason. They feel they are not being listened to when they lose. As a result, an us vs. them atmosphere is created on the team. Also, in the workplace, the losers frequently do not support the involved change; sometimes they even fight it. The worst possible scenario is when a manager or team leader is responsible for the final decision, with the rest of the members providing simply input or opinions.

Commitment to changes comes from universal process ownership. Such ownership, in turn, comes from consensual decision making. If a consensus cannot be reached on the nature of an improvement, it is usually because the team has not defined the problem accurately enough, because what they have defined is more so a symptom than the core problem itself. More questions need to be asked. If the core problem is, indeed, being addressed and consensus cannot be reached, one side is usually willing to try the other's solution first, with the understanding that if it does not work a change can be made. This ground rule could be considered team etiquette. However, it also has major implications for change in the corporate culture, for getting employees effectively involved in decision making.

9. Team members are not allowed to complain about a situation unless they are able to suggest a reasonable way to improve it.

This also could be considered a rule of team etiquette. It also, however, has important cultural ramifications for the organization as a whole. Complaining can consume entire team sessions and keep them from being productive, just as complainers in an office or on a shop floor can waste a lot of other employees' time while creating a negative atmosphere.

When teams are first formed and members realize they can speak freely, that they do, indeed, control the sessions, there is a tendency to use the sessions to vent frustrations against supervisors, other shifts, the benefits department, contractors, consultants, and so on. One hour per week, however, is not much time, and griping generally produces nothing positive except the temporary relief of the gripee's frustration. Therefore, meetings must be organized to utilize time effectively. The following agenda is suggested:

1. Review previously defined action steps of ongoing projects.
2. Record achievements.
3. Continue problem-solving effort with ongoing projects.
4. Define further action steps for ongoing projects.
5. Identify new priority projects to be worked on, when appropriate.
6. Identify new projects to be added to the list.

This is not to say that all griping is automatically eliminated. Especially during early, formative sessions it is necessary to allow members to ramble some. Gradually, however, the facilitator must become sterner, must begin enforcing the ground rules, focusing the team on problem definition and the generation of solutions. Most team members will appreciate and support this decision. The griping is nothing new, but the opportunity to make positive change is, and they will want to take advantage of it.

10. Teams have access to all company personnel to meet their information needs.

When a team is working on a project, the members identify whom they need information from and outline what they want. They then either assign someone to get that information or, through a team member or their facilitator, invite the source to attend the next meeting. Most specialists and managers appreciate this opportunity to talk openly with employees. For one thing, they consider such discussions an excellent vehicle for education. The team process is a learning exercise. Its strength is that team members themselves identify their learning needs. The willingness of an organization to make available the requested informational resources shows its continuing support for and realization of the value of the team process and its understanding that employees can make a valuable contribution if given access to the information they think important.

When someone is invited to attend as a resource, that person should be informed by voice or in writing ahead of time what the team wants to talk about, what the project is, and what information is being sought. This will allow the resource to prepare and to not feel that he or she is being set up or put on the spot. While access to all in-house personnel must be guaranteed, this ground rule should also apply to suppliers, customers, contractors, and other outside stakeholders whenever possible.

The president or facility manager is a favorite invitee to early team meetings. Members are curious to see if he or she will actually appear. When that person does show up, a very strong positive message is sent to the workforce. Most CEOs see such appearances as worthwhile expenditures of their time. After the initial visit, however, it must be explained to team members that project-related information can usually be supplied by someone other than the president or facility manager and that teams should work their way up the ladder in terms of resources rather than always starting at the top.

11. Team-suggested improvements must be justified by a cost-benefit analysis when possible or by a quality of working life rationale when not possible.

Requiring teams to justify projects is important for several reasons. First, it forces them to discover whether the proposed change is economically viable or not. It is a lot easier to say, "We need a new computer because this one is old and too slow," than it is to prove that the cost of a new computer will be offset in a reasonable period of time by savings and increased productivity. Second, it forces team members to think improvements through thoroughly.

The most accurate, comprehensive, long-term indicator of success in a quality process is improvement in the bottom line. With this in mind, teams should be forced to demonstrate the benefit of a new process control system, a new telephone system, or a new drill press in terms of dollars and cents. Again, this requirement makes the process an educational experience. It also encourages team members to learn what the organization's priorities are and why.

In some instances, it is difficult to define quantitatively the benefits of a project. An example would be improving a bathroom that has no heat, a dirty floor, no lock on the door, and graffiti all over the walls. It is a well-established fact, however, that workplace environment affects employee productivity. Thus, when teams list bathrooms, air quality, lack of access to food machines, too much human traffic moving through their area, leaking roofs, a lack of glare screens on computer terminals, or cold drafts as a priority, wise supervisors listen and give their support. The result frequently is improved department morale and a highly visible display of team success.

Cost-benefit analyses need not be detailed. Generally, they should provide a rough estimate of the problem's ongoing cost in terms of wasted employee time and materials; what the cost of the solution will be, especially if new equipment and labor hours are involved; and finally, an estimate of how long it will take the improvement to recoup what had been spent to make it.

12. Improvements/projects that might influence other parts of the operation must be agreed to by anyone else affected. Such stakeholders must also have a chance to contribute before implementation occurs.

A shift, department, or office supervisor decides that there must be a better way of completing a task. He or she asks a few questions, then figures out and institutes the improvement. Workers learn about the change after the fact and do not have the nerve to tell the supervisor that he or she has forgotten something important and, in the long run, has made the situation worse rather than better. A week or two later, other shifts, departments or offices, which the supervisor claims to have notified prior to the change, begin calling to say that they knew nothing about it and to ask if "Joe realizes how badly he has messed us up over here."

All of us are familiar with the above scenario. In order to avoid it, teams are required, when developing a solution or designing an innovation, to explain it to any who will be affected by it so that these stakeholders can challenge it or contribute from their own perspective. This ground rule forces the necessary integration of projects. It couples responsibility with authority.

A major problem with some quality improvement processes is that teams are encouraged to work on anything they want to without consulting others who will be affected by the improvement made. Frequently, the organization-wide, negative consequences of such unbridled freedom outweigh the positive local benefits. A second possibility is that two teams might begin working on the same project at the same time, thus wasting resources through duplication of effort.

The preceding is prevented from happening by this ground rule. It also tends to produce richer, more comprehensive, and more easily accepted solutions or improvements. The time wasted in gaining input and consensus is more than made up for by that saved during implementation. An additional benefit of such negotiations is that the involved stakeholder groups learn about other operations and about the nature of their relationships with each other.

Team members work on a problem until they achieve what is considered to be the best solution. The next step is to make a list of other functions that will be affected by the change. The strawman solution is then shared through informal discussion, in writing, by inviting representatives to a meeting, or by any combination thereof. If objections arise, alternatives to the original solution are worked out.

This ground rule, however, is the one most frequently broken. Once teams realize that they truly have been given the power to create change, they jealously guard projects, not wanting outsiders to interfere or to cause delays. It is the facilitator's responsibility, at this point, to force members to identify all stakeholders and to get their input. If the team has begun identifying or implementing action steps without doing so, the facilitator must make it stop, back up, and meet the requirement.

After completing several successful projects, participants understand that they do not lose, but, rather, gain by obeying this ground rule. Their results are better, enjoy more universal support, and are more loudly applauded.

13. Response to team questions/suggestions must be received within one week. The response to suggestions can be, "Yes, go ahead with it," "No, and this is why," with a reasonable explanation, or "Let us talk" with a date set.

Traditionally, workers with problems or ideas for improvements approach the supervisor individually or in small groups. The supervisor listens sympathetically, if there is time, then agrees and says that the matter will be attended to as soon as possible, or disagrees but does not have the time to explain why in detail, and promises to continue the conversation later. And that is that, until the idea or problem is brought up again by the worker several days, weeks, or months later, with much the same result. The supervisor is not uninterested. Rather, he or she is overloaded with ideas and problems pouring in continually from all directions to be dealt with during time not consumed by normal operational responsibilities.

One of the values of the team problem-solving and design approach is that it consolidates a majority of these ideas and problems into one list and prioritizes them through consensus, so that rather than 20 different employees with 20 different concerns, the supervisor must respond only to one group representing the whole and seeking help with one, two, or three projects at most. Another value is that team members no longer see the supervisor as a lap into which problems and ideas are dumped for resolution and implementation, but as a resource in their efforts to manage projects themselves.

The flip side of the coin is that while the pressure of numbers has decreased greatly, the pressure to respond has grown. This ground rule forces supervisors and other resources to take the time to think the issue through and to develop a useful response. It forces them to respond within one week. This does not mean that the desired information has to be delivered by that date or that the suggested improvement must be approved or disapproved. It means that the respondent must let team members know that their request has been received and is being addressed. The respondent must also set an acceptable date for delivery of the desired information or decision.

The purpose of this ground rule, then, is mainly to protect hourly teams from supervisors who cannot see that the teams will help, who are too busy to attend meetings, to make decisions, or to provide information. Most supervisors are quick to see the value of the team approach. A few, however, simply cannot accept the concept of hourly employees taking on more responsibility, making more decisions. Members of this latter group are usually too wise to balk openly. Their traditional strategy is to block projects until teams lose heart and quit. This ground rule makes such a maneuver more difficult.

14. When a ground rule is violated, the team facilitator will meet with the violator for resolution. If the problem is not overcome, the facilitator will report the violation to the head facilitator who coordinates the facilitator network. If the head facilitator cannot resolve the issue, he or she will carry it to the lead team.

The requirement of a one-week response time to team questions/suggestions helps expose those supervisors who want to block a QIP and the cultural change to which it leads. A favorite ploy of such people, as we have said, is to support team activities vocally while, at the same time, making sure that nothing gets accomplished. Such supervisors are not generally malicious. Rather, they are threatened or conscientious to a fault, firmly believing that if they do not personally control every aspect of the operation, bad things will happen.

If team members cannot gain the requested information, decision, or promise to attend a meeting, the team facilitator will get involved and speak directly with the desired resource, explaining the request and reminding that person of the ground rules. If the team facilitator is also unsuccessful, he or she will go to the head facilitator who will attempt to get thing moving. If the head facilitator also fails, that person has direct access to the head of the organization as a member of the lead team. This is the last resort, the hammer which all successful quality improvement processes require. A CEO interested in seeing the process succeed will make the necessary phone call immediately. The resource will cooperate; word of what has happened will spread quickly throughout the organization; and resistance will be forced further underground. If the CEO hesitates, however, and makes this a special situation, an exception to the rule, or suggests that the team move on to a new project, word will also spread quickly through the organization and the process will be dead.

Those opposed to change will test the team network's power in just this manner early on during the team building process. Everyone will be watching. This will be the CEO's first chance to demonstrate sincerity, to set the tone, and to begin moving the management culture in the right direction.

15. Task forces can be formed to deal with issues not being addressed by teams or to work on projects that include a lot of different functions, or to work on projects that will take a lot of time.

When a problem-solving or design team has to include stakeholders from other teams or departments in a project, there are four ways to get their input. The first and most common is to talk to them during working hours. The second is to send a representative to their function's team meeting. The third, conversely, is to invite them to the originating function's team meeting. The fourth, if the project requires a lot of work, is to create a task force to address the issues with each affected function sending its own representative or representatives.

Task forces can also be formed by the lead team or upper-level management around any project considered important that is not currently being addressed. QIP team members can be drafted to sit on these task forces, but not as an alternative to their team-related activities. Employees are allowed to sit on only one function team and one task force at any time, just as facilitators are allowed to facilitate only one team. It is important to interrupt the normal work schedule as little as possible.

As soon as the involved project is completed, all task forces, no matter what their origin, are dissolved.

Team Building as a Traumatic Experience

Another major reason teams fail is that they are not started correctly. Either too much control is exercised by those in charge, or too little. They never have a chance. They might last for months, or even years, but relatively little is accomplished.

In terms of our five musketeers, labor-management councils are the easiest to deal with. Labor-management councils are basically communications vehicles. Their role is not to make improvements, but to facilitate communication. Their charter, therefore, is clear, so that the main challenge is in getting members to talk, in developing the necessary level of trust on both sides. The desired results occur when participants are encouraged to be straightforward and to make sense, and when emotions are kept out of the exchange.

Self-directed work groups and circular organization decision-making boards also have a clear charter. These two types of teams are on the operating side of the equation, rather then the improvement side. Their role is to maintain and enhance productivity, again, through improved communication. They are relatively easy to start, once the decision is made.

Task forces and quality improvement teams, the heart of any successful quality improvement effort, are more difficult to nurture. They are responsible for almost all improvements made by employees, for almost all lower-level contributions to the long-range planning effort, and for the generation of employee commitment. Task forces and quality improvement teams, however, represent two sides of a previously discussed coin. Task forces represent the traditional, analytical side. A specific problem is picked out of the organization mess. It is defined and analyzed. Once a solution has been agreed on and recommended, the task force disbands. Quality improvement teams represent the synthetic side of the coin. Each individual team tries to identify, understand, and work on its own mess, its own comprehensive system of interrelated problems, while the team network, as a whole, addresses in an integrated manner the mess of the organization, in a participative, integrated, and ongoing manner.

Task forces, because they address specific problems, need access to problem clarification and problem-solving techniques that broaden perspective and help encourage creative thought. Often the problem has more than one interpretation. Often it has hazy boundaries or more than one part. A number of techniques have been developed which help define the essential nature of a problem and its boundaries. Three are brainstorming, problem setting and TJK.

A brainstorming exercise is basically an intellectual free-for-all. Participants throw out any ideas concerning the true nature of the problem that come into their heads, no matter how bizarre, building on each other's contributions, not making judgments. When no more ideas are forthcoming, the contributions are grouped according to similarities. Then the groupings are ranked by vote.

In a problem setting exercise, after the initial version of the problem is written on a board or flipchart, three lists are made based on participant input. The first is a list of problem producers. The second is a list of potential consequences if the problem is not solved. The third is of changes that need to occur. The relationships between items in the three lists are explored, and lines are drawn to show linkages. Finally, these relationships are prioritized according to their relative importance, thus, helping to clarify the problem.

In TJK, after the initial version of the problem has been written down, participants list problem-related facts. The lists are exchanged. Then a single fact from one list is read aloud and also written down. Other participants read related facts from their lists, which are added to the first. Eventually, a new list is begun with a new lead fact, and so on until no more facts are left. The next step, after discussion, is to begin consolidating the various lists when possible.

These three techniques provide a means of exploring different interpretations of a problem, of clarifying the problem, and of leading participants toward consensus. At least brainstorming and TJK can be used for problem solution as well, along with the nominal group technique, the dialectic, which is especially good when conflicting solutions are proposed, and swapping, which is good in the same situation. Elaboration on these techniques and on their value can be found in *Problem Solving for Results* (St. Lucie Press, 1997).

Quality improvement teams, as part of a systemic quality effort, must begin by defining the role that the function they represent plays in the larger environment of which that function is a part. Instead of focusing on one problem, therefore, they must begin by attempting to identify the network of problems affecting their operation. Two systems-based techniques that meet this challenge are the search conference and the modified idealized design.

The search conference is especially useful with hourly teams because it begins with the necessary overview, then works its way down to their immediate concerns in a straightforward, non-threatening, no-tricks manner. It also, if used correctly, immediately puts participants in charge. Finally, it is not difficult to learn, so that in-house people can quickly be trained by the consultant to lead the exercise, thus multiplying the number of teams that can be started simultaneously. The search conference is based on a set of questions participants are asked to respond to, their answers being noted on sheets of flipchart paper and taped to the conference room wall for ongoing reference. The questions are:

1. What trends in the United States and your community are most affecting your quality of life?
2. What trends in your company as a whole are affecting your quality of working life?
3. What trends in your unit (division, department, office) are affecting your ability to do your job the way you want to do it?
4. What are your responsibilities (briefly)?
5. What changes would you suggest (in products, production processes, management systems, the work environment) that would help you better meet your responsibilities in your unit?

Lists are made for all of these. When participants have completed their list of changes, the consultant takes them back over company and unit trends, and, finally, over responsibilities, asking questions to see if anything has been missed in terms of possible improvements. The next step is to divide the changes into three categories:

1. Those the team can address without input of anyone outside the boundaries of its function.
2. Those that need input from other functions.
3. Those that need input from the lead team owing to cost or because they necessitate changes in organization policy.

The final step in the exercise is to prioritize the changes, usually starting with those which can be accomplished quickest with minimal expense. Most such changes, of course, come from the first category.

The modified idealized design technique is a watered-down version of the idealized design technique core to the planning phase of any systemic quality improvement process. It can be used by task forces, but is most effective when introduced with management-level quality improvement teams addressing entire systems. Team participants are asked to compile three lists:

1. External stakeholders.
2. Technical systems they work with and by which they are affected.
3. Key management systems by which they are affected.

The managers then pick the system most in need of improvement, reach agreement on what its role should be ideally, define the characteristics it should have ideally, and, finally, identify and prioritize action steps necessary to move toward these characteristics. The modified idealized design technique is not as easy to learn as the search conference and, therefore, should be run by an experienced consultant. The main challenge is to keep participants from fixating on the current situation, instead of working back toward it from the ideal.

By way of summary, then, in terms of teams, we now, hopefully, understand the five different types at our disposal and how they fit together and complement each other. We understand which ones are on the operating side of a business and which are on the improvement side. We understand which should be management controlled and which should be employee controlled.

The two types of teams mentioned that are most critical to a successful QIP are the quality improvement network teams and the management directed task forces. We have talked about the importance of starting these two correctly and have suggested techniques that can be used to do so, techniques which help members organize their ideas, but which do not take control of the process away from them. Finally, we have discussed the ground rules necessary to their success.

Topics for Discussion

1. What problems has your team building effort run into? How have your dealt with them?
2. Why is it necessary that everyone agree to the ground rules up-front?
3. Which ground rules are most important in terms of employee empowerment?

4. Which ground rules are most important in terms of protecting team projects?
5. The institution of which ground rules would make the biggest difference in your organization's team building effort?
6. From a systems perspective, what are the major differences between the three problem solving and the two start-up techniques discussed?

5 Five Phases Plus Two: Part II

After Reading This Chapter You Should Know

- The role of the facilitator.
- The way training needs should be identified and met.
- Why most attempts at the introduction of statistical techniques fail.
- Why the interactive planning paradigm meshes most effectively with quality improvement processes.
- The only accurate way to measure a quality improvement process' success.
- The reward system most likely to encourage employees to continue improving their efforts.
- How to get middle management buy-in.

Training as a Participative Challenge

The training phase starts soon after the first teams are in place and follows the team building phase through the organization. It follows because initial facilitator training is done on the job and because team members identify a large number of their own training needs.

Teams, as we have said, have facilitators and not leaders. The facilitator's job is basically to make sure that the ground rules are obeyed, to help keep meetings on track, to function as a resource or resource generator, and to help coordinate his or her team's activities with those of the other teams in the network. The facilitator cannot be any team member's boss. Employees must realize through the team process that they can and should begin thinking for themselves, that the company can no longer afford for them to wait to be told what to do. This change in workplace attitude is nearly impossible if a boss is given the same position of control in the team meeting that he or she has on the office or shop floor.

The facilitator can be either an hourly person (for hourly teams) or a manager (for managerial teams) but should be from another part of the organization so that personal opinions about the best solution to the problem being worked on do not confuse his or her role. The facilitator, as we have said, is initially trained on the job. After a brief familiarization session concerning the team building process, each facilitator sits in on team meetings led by a consultant. Eventually, the new facilitator is given the lead, with the consultant observing and supporting. Once several facilitators have been trained in this manner and begin communicating, they help to identify other training needs; they set up their own training sessions; they assist each other with insights.

Team members also receive almost no initial training. They are taught a very simple problem-solving technique, mainly by applying it to the first actual improvements they decide to work on. Further training needs are identified, at least in part by the team itself. A majority of these needs are technical rather than a group process. They fall roughly into four categories:

1. Additional job training for new employees who were not adequately trained when their responsibilities were first assigned.
2. Cross-training once employees have become proficient in their own areas of expertise.
3. More training on new equipment now that employees have become familiar enough with it to identify what their weaknesses are.
4. Training concerning the organization-wide production process so that employees can better understand how they fit into it.

The Introduction of Measurement Techniques

In a production unit of a thousand or so, educating quality department staff to the nuances of the five-phase process, familiarizing everyone, creating and integrating the necessary network of teams, forming and integrating the necessary network of facilitators, and providing all the critical and requested training will probably take at least a year. By the end of this period, however, even though all of the teams might not yet be operational, a critical mass should have materialized or be well on its way to materializing.

In a QIP, the concept of critical mass relates to the point at which enough people understand and believe in the value of the process that it cannot be stopped except by executive edict. It refers to the time when the process takes on a life of its own. Workers and managers rather than consultants now conduct the ongoing familiarization effort. Trained in-house staff take the new teams through their initial problem identification and prioritization exercise. Training needs have been well-defined and met. The teams have generated several hundred useful improvements and have been continually applauded for their efforts by the lead team.

Also, by this time, many teams have solved all the simple problems affecting their productivity and are realizing that additional improvements will be more difficult. They are beginning to understand the complexity of the operation. It is at this point, then, that measurement techniques can be effectively introduced as a potential aid to their continuing efforts.

The stage is set. The trick, however, is to give the teams ownership, to let *them* decide that what is being offered is a good idea and should be spread throughout the operation. The basic techniques involved—histograms, run charts, control charts, flow charts, Pareto, the fishbone, and scatter charts—are not that difficult to teach or to use. The hard part, traditionally, has been to get employees to *want* to use them.

There are three possible approaches. The first is coercion: "Learn how to use these or start looking for another job." However, when employees are forced to implement something they do not understand the need for, they tend to make a lot of mistakes.

The second approach is to educate workers to the value of measurement techniques. Conscientious and ambitious employees will listen and will make a sincere effort to implement what they have learned. Those just putting in their hours, however, will see the additional time and effort requirement as another unwelcome imposition and will take shortcuts.

The third approach is to encourage the employees themselves to discover the value of the techniques. Chances for successful implementation increase greatly when team members begin saying things like, "We need to understand the entire pattern of paper flow in this office before we can successfully resolve bottlenecks," or "The only way we are going to find out what is wrong with this machine is to keep an accurate record of variations in run quality." When this last approach is taken, managers and trainers are no longer totally responsible for convincing workers of the technique's value. Also, rather than managers having to make sure that everyone enters their data points correctly, the employees themselves willingly accept this responsibility.

They keep tabs on, encourage, and assist each other because *they* have ownership.

The third scenario is obviously superior, but how do we achieve it? We do so primarily by developing a strong sense of employee commitment to improved quality in products, manufacturing processes, management systems, and the work environment. But how do we engender this required commitment? The answer is that we foster it through the familiarization, vehicle emplacement, and training phases that precede the introduction of measurement techniques.

Strategic Planning Reborn

In order for QIP organizers, familiarizers, team facilitators, trainers, and team members to succeed, they must have a framework built on clearly defined organization objectives into which to fit their integrated efforts. This framework, or overview, traditionally has been created through an annual strategic planning exercise. Recently, however, strategic planning has fallen on hard times. It is no longer considered a critical function in many corporations. The future of such organizations is now being decided by the president or CEO and a few confidantes, their decisions based on a mix of financial considerations, their own observations, and gut instinct.

The reason for the recent disenchantment with strategic planning is twofold. First, the ever increasing amount of turbulence in the environment makes the ability to react quickly to shifting environmental circumstances more important than the ability to generate a long-term plan. Second, even if our orientation were long term, strategic planning, as it has been practiced during the 1970s and 1980s, does not work.

According to Russell Ackoff's article "The Corporate Rain Dance," which appeared in the Winter 1977 issue of *The Wharton Magazine*, strategic planning during this period was carried out strictly in either a bottom-up or top-down fashion. In the bottom-up approach, word was sent out annually for low-level unit heads to put together and prioritize their budget requests, then to submit these requests to their superiors, who were to do the same thing, and so on until final lists reached the top for scrutiny and further deletions.

There were many weaknesses of this approach. First, the budget estimates were never accurate. Management expected requests to be inflated and automatically cut the amount allowed. In order to get what they actually needed, therefore, departments doubled or tripled their projected requirements. Second, decisions made at the various levels concerning the lists coming in from the next lowest echelon were frequently influenced by corporate politics or were based on insufficient information. Third, very little true integration of departmental priorities occurred, as units were forced by the process to compete for resources.

The top-down approach began with the formation of a strategic planning department. This department was separate from the rest of the company and reported directly to the CEO. Its role was to do ongoing research and to develop models and projections concerning technology, the market, the competition, raw materials, capital, the political situation, and so on. It was to consolidate this information into recommendations from which the CEO could devise a set of overall long-term corporate objectives. These objectives would then be passed down so that each department and level could define the changes and activities in its area necessary to meet them.

The problems with the top-down approach were also many. To begin with, owing to the continually increasing number of relevant variables in the organization's environment, it was impossible to identify and incorporate all of the important ones. Even if one did succeed in this Herculean task, however, owing to the increasing amount of turbulence in the relationships between the involved variables, suppositions drawn that were correct one day might not be correct the next day, or the next week, or the next month. By then corporate and departmental objectives had been defined and units had begun planning so that word had to be sent down to modify. Yet while the necessary adjustments were being made, a second critical variable began shifting, or was threatening to shift, and so on.

Corporate objectives defined in this top-down manner also frequently failed to take into account the situation at lower levels. Impossible demands were made that eventually had to be modified, thus affecting the organization's ability to meet its original objectives. Finally, the necessary degree of integration between departmental efforts rarely occurred.

Strategic planning is important, not only to quality improvement, but also to the long-term health of any organization. Asian and European firms do it effectively and have used the results to increase their share of world markets. The problem, therefore, is not that strategic planning does not work, but that the paradigms that we have been using in the United States are not right for our situation.

In order to be more successful, we need an approach that helps to replace or to combine our short-term orientation with a long-term one; discourages in-house competition for resources; helps to generate consensus on priorities; encourages the necessary integration; effectively reads and reacts to the organization's increasingly turbulent environment; and does not impose unrealistic demands on those required to translate corporate objectives, as defined, into reality. In other words, we need an approach to strategic planning that is systemic.

The interactive planning paradigm developed by Ackoff and described in his book *Creating the Corporate Future* meets all of the above criteria and is, indeed, systemic. It also enjoys a comfortable fit with and is fueled by the other four phases of a comprehensive QIP. Interactive planning has the same key characteristics that all systemic efforts share. First, it is *participative.* All employees are expected to contribute to and learn from the process. This helps foster the necessary consensus and integration. Also, it helps avoid surprises and unreasonable demands, enlists the aid of the entire work force in spotting shifts in key variables, and greatly enriches the quality of input.

An obvious objection to such extensive participation is that it draws out the planning process. The time factor, however, becomes irrelevant when we learn that the second critical characteristic of interactive planning is its *continuous* nature. When an organization chooses this paradigm, planning is no longer seen as a time-consuming annual exercise. Rather, it becomes part of the normal daily or weekly routine. This characteristic allows the plan to evolve and to improve as shifts in key environmental variables occur. Short-term needs also can now be taken into account, but are framed in terms of realistic long-term objectives.

The third key characteristic of interactive planning is that it is holistic, addressing every part and level of the organization simultaneously and making sure that what is generated is well-integrated. And, of course, interactive planning allows constant feedback, learning, and self-improvement, the fourth characteristic.

Interactive planning, like comprehensive quality improvement efforts, has five overlapping stages. These are formulation of the mess; ends planning; means planning; resource planning; and implementation and control.

Formulation of the Mess

A mess is defined by Ackoff in "The Corporate Rain Dance" as a system of problems. During this stage the future of an organization is charted if no changes are made and if the environment remains stable. The stage includes a comprehensive series of systems analyses covering both the organization's internal operations and its environment, as well as traditional reference projections. It synthesizes all of the preceding into a reference scenario that more graphically and creatively shows employees where they are now, where they are headed, and the types of changes necessary if the organization wants to improve on or even maintain its current level of success. While some of the previous exercises, such as the reference projections, need to be conducted by professionals, others can be contributed to by QIP teams.

Ends Planning

During this stage participants define an ideal. They are told that their company, mill, or department has been destroyed and that they are responsible for totally rebuilding it. The charge, however, is to design what *should* exist ideally, rather than simply to improve on what originally existed. The only three constraints to this exercise are that the systems designed must be technologically feasible, financially reasonable, and capable of adapting to future environmental change.

Everyone is involved. Executives begin with the organization as a whole, idealizing its overall mission, objectives, and goals. Supervisors begin with management systems. Hourly workers begin with their manufacturing processes and work environment. As the designs are completed, they are integrated with those of bordering units and reworked in terms of this broadened perspective.

Once completed, the final, comprehensive idealization gives the company an agreed upon target to aim for so that the limited number of changes a yearly budget will allow makes sense as part of a long-term, well-integrated whole. At the same time, the exercise result continues to evolve and improve.

Again, the organization-wide team network provides a perfect vehicle for an idealization exercise. The teams in a QIP progress naturally from problem solving to design challenges, from quick fixes such as the marking of pedestrian lanes in a warehouse to reorganizing the warehouse so that there is less traffic overall. When management teams and, eventually, hourly teams begin addressing design challenges, they should be taught the idealization technique, so that this stage of interactive planning is already occurring piecemeal throughout the facility.

Means Planning

During this stage projects that will help move the organization from its current mess toward the ideal are defined and prioritized. The QIP teams can play a key role in both definition and prioritization once overall organization objectives, once the ideal has been identified.

Resource Planning

The teams can contribute heavily to both the identification of available resources and allocation decisions.

Implementation and Control

The teams can assist in carrying out action steps. They can also generate the information necessary for proper control and function as a channel for the upward and downward flow of this information.

Concerning the introduction of strategic planning, timing is critical. As mentioned earlier, hourly and supervisory teams must be allowed to gain the necessary degree of process and team project ownership before upper-level management asks them for help or otherwise directly orients their efforts. Planners must wait at least until a critical mass has developed to begin using the quality process vehicle to achieve their own ends. Otherwise, they break a ground rule. They can ask individual employees, shifts, or departments for assistance, they can form task forces, but they cannot use the QIP teams.

The other side of the coin is that the CEO or unit manager, and his or her direct reports, should start their own idealization exercise as soon as possible. The objectives defined at this level will indirectly impact every other team's perspective and activities by providing the previously mentioned organization-wide framework.

Measurement That Makes Sense

Quality improvement processes, of course, also need a way to measure their success. Most current approaches to measurement, however, produce results that are, at best, dubious. The reason for their failure is that, once again, they do not focus on to the degree to which overall, organization-wide effectiveness has been improved. Rather, they focus on identifying improved efficiencies in parts of the organization. In terms of the big picture, therefore, the results tell only part of the story. They can be used as showpieces. They can be used to give teams and the work force recognition for their efforts, but they should not be taken too seriously.

The problem is that the measurement of improved efficiency is normally done on a project by project basis. An accounting team saves the department $20,000 by developing a new technique. A receptionist team project cuts 25 minutes from the hospital patient intake process by condensing five forms into one. All these dollars and hours saved are then added up, the total being presented as proof that the quality process is succeeding.

Sound familiar? What we end up with once again, when using this analytical, piecemeal approach, is an aggregate of individual success stories which might support each other in terms of overall organization benefit, but which, more frequently, do not. Increased efficiencies in one part of the organization might actually create unrecognized or ignored inefficiencies elsewhere that offset the original savings. No one really knows. And no one, with the exception of top-level management, usually cares, because the measurement system has been designed to focus attention on the success of individual units, ignoring the fact that the whole might still suffer.

If a quality improvement process is designed to be systemic, the measurement system must also be systemic. The focus in a systems scenario is not only on improving the organization parts. The focus is, even more so, on improving the interactions between these parts, on optimizing the results of their interdependencies across the entire organization.

In the systems scenario no individual success stands alone. Very few projects do not affect other departments. All improvements, no matter how small, ripple outward. We do not understand their total effect until these ripples eventually break up against the containing environment.

As a result, in the systems scenario, it becomes obvious that the only true quantitative measure of a successful quality process is the *bottom line.* This is the only indicator that captures the holistic effect of the total gamut of process-driven improvements. Other variables, of course, affect this figure. But the bottom line is as close as we are going to get to reality.

Teams, therefore, should be praised for individual successes, but team members should also understand the need for their successes to fit into a bigger picture. They should understand that a willingness to sacrifice or ignore the needs of other parts of the operation in order to look good is not what it is all about.

A second, non-quantitative but unfailing measure of success in quality improvement efforts from a systems perspective, is improvement in employee morale. If an operation becomes more productive it does so because the potential of the work force is being better utilized, because individuals and units are cooperating and supporting each other more effectively. The positive impact of this more respectful, more cooperative, more productive, and profitable environment is normally reflected by a rising level of employee morale.

Rewards as the Key to Long-Term Commitment

Any description of a systemic approach to quality improvement must eventually hone in on the reward system associated with the process. Unless the reward system is set up correctly, no matter how well the other pieces are crafted, the process inevitably becomes a sham. Two types of rewards must be discussed. The first is long-term rewards that center on salaries, bonuses, and benefits. The second is short-term rewards or encouragements developed to reinforce team efforts and successes.

Most long-term reward systems in place today (normally on the operational side of things) negate the possibility of gaining the level of employee buy-in and commitment necessary to a successful quality improvement process. They pit employees and departments against each other under the false assumption that competition, that win-lose situations bring forth our best efforts. The essence of a systemic approach to quality improvement is *cooperation.* That is the only way the pieces can be integrated to the necessary degree. The company's normal reward system, therefore, must also encourage cooperation. It must encourage employees to help, rather than to try to beat each other. At the same time, it must allow employees on all levels to benefit directly from their own efforts, but never at the expense of the organization as a whole.

This brings us back to the bottom line. When we reward individuals or individual teams financially for their contributions, whether the effect of these contributions can be seen on the bottom line or not, we are again thinking incrementally. We are ignoring the fact that others might have contributed. We are ignoring the fact that a wide range of functions must cooperate for this improvement to have the desired impact. We are ignoring the fact that this improvement might trigger negative consequences elsewhere in the organization.

The only long-term reward system that will help pull quality improvement efforts together in the necessary manner is, as we have said, one tied to the overall bottom line. Lincoln Electric has been doing it for years. Let employees see the benefits of their efforts in their paycheck and bonuses, but as a team of the whole, not as individuals. Think synthetically. Expect everyone to be a good performer.

"But what about employees who try to get by doing as little as possible?" we hear. The answer is that in a real team atmosphere, where the cooperative ethic holds sway, where the necessary ground rules have begun to seep from the quality effort into the ongoing, daily operation, where the workforce has truly been empowered, the employees will take care of such issues themselves, usually in a more direct and effective manner than management.

Short-term rewards or encouragements developed to stimulate quality team performance can take many forms ranging from a simple but sincere "Good job!" offered by a manager or another team, to dinners, days off, certificates, picnics, movie passes, or giving employees a chance to show off what they have accomplished to the community or to other companies.

The requirements concerning such encouragements are few and simple. First, again, they should not generate a win-lose environment. Everyone should be able to earn them upon reaching a certain level of performance which does not continue to go up. Second, they should not become the focus of the effort. Third, they should not be monetary in nature, or too valuable. Save the valuable stuff for the long-term reward arena.

Caught in the Middle

As a final point in this section, we should discuss the quandary of management. Middle management has been defined as the major source of resistance to the implementation of successful quality improvement processes. These people seem to be the bad guys. From a systems perspective, however, they resist change not because they are against it, but because they are suspicious of where it is leading. Most change efforts have failed to spell out the future of management, especially the middle level, except to say that there will be less of it. Most do not understand that managers are critical to a successful QIP and, if approached in the right way, become the process' strongest advocates and most productive participants.

It is with good reason that, when first introduced, many middle managers, as we have said, see a quality improvement process as a threat. They fear that it will eventually eliminate their jobs as the number of management levels is decreased to improve communication. They fear that their decision-making responsibilities will be lost to lower-level employees leaving them with nothing to do. They fear that they will ultimately be saddled with the blame for poorly thought out changes designed by lower-level employees.

All too frequently, as corporations rush to join the quest for improved quality, these fears prove well-founded. The problem is not the corporation's ultimate objective. Improved quality is a must if a competitive status is to be maintained. The problem, instead, is with the approach. It is frequently not well enough thought out. It is not systemic. It has good parts, but lacks the necessary comprehensive whole.

The primary goal of a systemic quality improvement process is to increase employee involvement. The most important initial target is usually defined as the lower level, hourly employees. It is natural for top-level executives who champion quality improvement to think this way because they enjoy little interaction with hourly workers. They assume, and usually correctly, that the hourly worker is not taking enough responsibility, does not have enough authority, is not making enough decisions in his or her own area of expertise.

At the same time executives do interact, frequently on a daily basis, with middle managers. They know that middle managers spend most of their time solving problems and making decisions. Efforts to increase involvement and commitment at this level, therefore, are not as critical.

It is this misinterpretation of middle management reality and not the middle managers themselves that is the real cause of the breakdown in so many quality improvement efforts. The truth is that most middle managers ultimately have as little control over their time and and activities as hourly workers. On top of that, they are frustrated because their responsibilities and authority are not clearly defined. Decisions supposedly theirs to make are too often usurped by bosses.

The purpose of this section, then, is two-fold. First, it is to help clarify the role middle managers must play in terms of hourly employees if a quality improvement process is to produce the desired results. Second, and more important, it is to emphasize the opportunities such processes must offer middle managers as well if success is to be achieved.

Setting the Stage

Combining managers and hourly workers on newly formed ongoing teams (not task forces), as we have said, is a mistake. Either the managers take control so that it is business as usual and the necessary sense of team ownership is not generated in hourly members, or the managers say little in their effort not to dominate, but instead spend this wasted time worrying about all the things they should be doing while they listen to the hourly employees muddle through.

Hourly workers should have their own teams. Eventually, they can decide to invite managers to join, but they should begin without them. The hourly team members should identify, prioritize, and work on the projects most important to them. Thus, process ownership and commitment are generated. They should be able to bring managers to specific meetings to provide information and advisement on specific issues. This issue is addressed by ground rule number 10. At the same time, they should be required to get the input and the approval of any manager who will be impacted by a team project before it is implemented. This issue is addressed by ground rule number 12.

In terms of hourly teams, then, managers should:

1. Show positive interest in and support for activities.

Ignoring the teams, being too busy to read meeting minutes sent to them, to answer questions, or to provide requested decisions are behaviors that will cause team efforts to fail and team members to lose their willingness to participate.

2. Exercise patience.

The teams are going to make mistakes. They are going to take longer to define solutions than more experienced managers would. The latter, however, must realize the process to be, as much as anything else, a learning experience for lower-level employees. The employees are developing a more comprehensive understanding of the systems in their own production areas as well as of relationships with other production areas, especially those that border theirs. Managers should understand the value of this learning experiences and facilitate it, rather than pushing constantly for rapid results.

3. Respect team-defined project priorities.

All employees want to work on the improvements most important to themselves first. Only then will they pay serious attention to the concerns of others. More often than expected, hourly team priorities mirror those of the managers. When they do not, however, managers should not insist on change.

In primary manufacturing situations initial team priorities are frequently safety related. For the hourly employee who must face safety hazards on a daily basis such improvements are critical. In situations where employees work in offices, initial priorities are frequently environmental. While a draft, continuous glare, or chairs with backs curved the wrong way might seem relatively unimportant to managers, research has proven such discomforts to strongly affect morale and, indirectly at least, productivity.

4. Never try to tell hourly teams what to work on.

The temptation to do so is great, but, as we have said, no surer way exists to kill team spirit. Team members' expectations begin rising after the first few successes. The members start thinking, "Management really does want our input." When a manager, therefore, turns around at this point and says, in effect, "Okay, folks, now that I have let you play for a while I am going to tell you what to work on," the process loses credibility and the company loses the opportunity to more effectively utilize its employees' expertise and energy.

An acceptable alternative is for a manager to wait until the hourly teams have dealt with all projects of immediate importance to their members, then to submit a list of issues to be worked on if the teams choose. Members, because they have gained the necessary sense of process ownership and self-confidence by this time, will normally welcome such a list and attack it eagerly.

5. Never punish hourly employees for team-related activities.

Managers should carry complaints to the team facilitator for resolution. If not satisfied at this level, managers should move next to the head of the facilitator network. Team members who feel they are being punished for team related activities should also bring their problems to their facilitator. Such punishment can be subtle. It can be limited to verbal abuse or sarcastic comments. If it is ongoing, however, team members should act.

Beyond Holding Hands

If middle management does not meet the preceding requirements concerning hourly teams, chances of a quality improvement effort succeeding will be limited. Simply telling middle managers to meet them, however, does not usually work. They are no different from anyone else. They are being affected by this process and, therefore, want a say as much as the hourly and professional workers do. They need to be shown by top-level management that their input is also valued and trusted.

It is necessary, therefore, to find a way to quickly give managers as well as hourly workers process ownership. The best vehicle is again the team. Middle managers should have their own teams governed by the same ground rules as hourly teams. Members should have the right to identify their own projects and priorities. They should have access to desired information and resources and should be required to gain input and agreement from anyone affected, including hourly workers, before implementing an improvement.

Teams of managers, again, as we have said, can be formed by department, by level, or by both. Level managerial teams usually begin by attempting to clarify their job responsibilities and authority and their relationship to superiors. By picking responsibilities and authority as their number one priority, level managerial teams often bring the quality improvement process to a crossroads. They force corporate executives to demonstrate process support with actions as well as words. If the corporation's vice presidents and CEO balk at discussing their relationships with middle-level managers, if they are too busy, or indicate that improvement efforts should be kept at lower levels, the process is dead. Even if lower-level managers see the value of a more open, participatory style in dealing with subordinates, they will be reticent to change for fear of being labeled eccentric or radical by superiors. The process might continue, but the necessary degree of commitment will not materialize.

In summation, then, middle managers are key to a successful quality improvement process. If they do not support hourly team efforts, the process will die. If they are not given a vehicle of their own which allows them to make changes they consider important, rather than functioning simply as resources and facilitators to the efforts of others, the process will die. If they are rebuffed when they request a clarification and perhaps a rethinking of their relationship with senior management, the process will die.

But if middle managers are allowed to effectively utilize their years of experience to improve products, manufacturing processes, management systems, and the work environment, the process and the organization will flourish. Middle managers are the link. They are the puzzle piece that ties everything else together and integrates the whole. This makes the meeting of their process related needs a number one priority in any successful quality improvement effort.

The Laid-Back Lead Team

Senior-level management, in the systems approach to quality improvement, has to rewrite the book. It has to get things started, then step back and let others lead the charge. In our culture of aggressive go-getters this is, to say the least, difficult. A handful of CEOs have managed to do it—Peterson of Ford, Stayer of Johnsonville Foods, Teerlink of Harley Davidson, for example, but most prefer to stay out in front; most can see no other way.

In the systems approach top-level management does become heavily involved in the familiarization phase in order to show its support. Once the team building phase begins, however, it gets out of the way, focuses on providing support rather than on continued domination.

The most important of the lead team's responsibilities, as we have said, is to begin and lead the ongoing interactive planning process. The lead team must also function as the previously-mentioned process hammer, helping ensure that the ground rules are obeyed without exception. The lead team must contribute to the ongoing familiarization process. It must function as an informational resource to all teams in the network, but also important, lead team members must function as a spiritual resource to process participants, listening, continually encouraging and stroking without taking over.

The lead team must also help ensure that quality team and task force efforts are integrated. The lead team is a focal point of the enhanced organization-wide communications system that results from activities of the team network. The network's tentacles reach into every corner of the operation. The lead team must tap into these tentacles and use them to help tie everything together.

In essence, there is no change in authority patterns, at least initially, on the operating side. On the quality improvement side of the equation, however, senior-level management, the lead team, must be there, must be deeply involved, but in a laid-back rather than a take-charge manner.

The Systems Approach

So that is it, that is the systems approach to quality improvement, that is why the concepts of systems, quality, and, in the long run, organization development cannot be separated. And it is all based on a one-liner with no commas, one that reads, *a whole is more than the sum of its parts.*

The systems approach to management is a straightforward concept and a necessary ingredient to successful organization change. When we think systemically, we must, as always, identify the key parts of the organization; but, not as always, we must then move beyond these parts and focus on their interactions, defining them in terms of function, processes, and structure while never losing sight of the whole they belong to, never losing sight of that whole's unique characteristics.

When our concern is improving quality, we need to build our effort from this same perspective. The parts of the quality effort must create a comprehensive whole that effectively integrates function, processes, and structure and that encourages the participation, holism, ongoingness, and feedback necessary to comprehensive improvement.

The parts of a quality process are relatively easy to identify—familiarization, team building, group process skills training, introduction to statistical measurement techniques useful in the workplace, and long-range planning. Designing the necessary characteristics into these parts, however, so that together they create the desired whole with the desired characteristics, takes an advanced degree of understanding. It also takes a realistic means of measuring success and a reward system that reflects reality, encourages employee cooperation, and encourages a focus on organization-wide effectiveness rather than on just efficiency.

The systems approach to quality improvement makes sense, so the hard part will not be learning it. Instead, the hard part will be implementing it, overcoming old, deeply entrenched habits.

But then, as we all know, real change, especially in the way people think, has never come easily.

Topics for Discussion

1. Why in the systems approach are there no team leaders?
2. Why do companies have so much trouble introducing statistical measurement techniques?
3. What differences exist between your company's approach and the interactive approach to long-range planning?
4. Why do companies spend so much time trying to develop ways to accurately measure the success of their quality improvement effort?
5. Does your company's quality improvement effort reward system encourage cooperation or competition?
6. From a systems perspective, what is the best way to get middle managers to accept their role in the process?

6 More Than One Way to Skin the Quality Improvement Cat?

After Reading This Chapter You Should Know

- The key characteristics of and steps in the Baldrige Award-winning Ames Rubber Company model.
- The key characteristics of and steps in the International Paper systems model.
- The strengths and weaknesses of the Baldrige model in real world situations.
- The strengths and weaknesses of the systems model in real world situations.
- Why and how we are moving gradually toward the systems model.

Comparing the Systems And Baldrige Models

The systems approach to quality improvement is obviously not the only one available. In fact, it is not even the most popular. In spite of the fact that, when used, it has produced excellent results, most companies find it too radical. They prefer a diluted version. The most popular current alternative is what we call the Baldrige model because most of the companies currently winning the yearly, highly prestigious, government-sponsored Malcolm Baldrige Award for their quality improvement efforts have adopted it.

In this chapter we want to compare the previously defined systems and the Baldrige models. In order to do so, we will introduce two highly successful efforts at quality improvement in two known companies, one systemic in all respects, the other partially systemic at best. We will discuss their differences, then evaluate them by systems criteria. We will ask whether each of the models includes a systemic whole, whether it includes all the necessary pieces—familiarization, long-range planning, team building, groups process skills training, introduction to statistical measurement techniques and mapping techniques useful in the workplace, a way to effectively measure the results of the organization quality improvement effort, an appropriate reward system—and whether these phases are integrated in the necessary manner.

We will then try to discover if the two models incorporate the necessary systemic characteristics, if they are participative, wholistic, ongoing, provide continual feedback—and if these characteristics are possessed by each of the key pieces. Finally, from a developmental perspective, we will ask to what degree the two models enhance real employee control over their work lives and to what degree they enhance employee *empowerment.*

The first of these models is the one people involved in the quality improvement movement are most familiar with. We call it the partial empowerment model. The Ames Rubber Corporation is an organization that has improved performance and gained national recognition by implementing it. We will use this corporation as an example.

Ames Rubber is an extremely well run company that won the Baldrige Award for Quality in 1993. The company's change effort began in 1987 when Joel Marvil, CEO and Chairman of the Board, visited one of the company's major customers, Xerox, which was in the process of rescuing and revitalizing itself through a TQM process. Xerox was also a Baldrige Award winner. Joel liked what he saw and carried the concept home. The next step was to send a group of executives for training in Xerox's approach.

Once this training had been completed, Joel and six members of the executive committee met over the course of the next year to define the 25 strategic tasks they considered necessary for the development of a total quality strategy and the improvement of Ames Rubber Corporation's performance. These tasks were:

- Define expectations.
- Define the cultural changes required.

- Create a vision statement.
- Create a mission statement.
- Name a strategy.
- Create a strategy.
- Answer the question, "Why total quality?"
- Create a quality policy.
- Define strategic quality goals.
- Develop an orientation plan.
- Develop a training rollout plan.
- Identify barriers to rollout.
- Develop a team management plan.
- Develop a system for recognition and reward.
- Develop a communications model.
- Develop a way to measure the cost of quality.
- Develop a vehicle for competitive benchmarking.
- Develop a vehicle for coaching and inspecting.
- Develop an approach for modifying management behavior.
- Identify top-level management's role.
- Define overall roles and responsibilities.
- Define ways to introduce statistical tools.
- Develop a follow-up training plan.
- Develop a measurement plan for process effectiveness.
- Develop a model for strategic review.

The next challenge was to disseminate the strategy developed during these sessions and to deliver total quality training to the work force. A cascade training program was designed. Starting with the executive committee, each of the four management levels in the organization was responsible for educating the level below and for monitoring its progress. The concept of TQM was introduced, as well as the skills required for effective problem solving, team building, and benchmarking. This training exercise filled roughly the second year, at the end of which time all six hundred employees of Ames Rubber had been reached.

Actual changes began occurring when the CEO and his six person executive team began rethinking corporate policy and redesigning the organization from top to bottom. When the rethinking and redesign had been completed on paper, this group began implementing the changes required at their level. They then passed the new design, the new policies, and the responsibility for implementation on down to the next level, putting the managers at this level in charge of leading the change process organization-wide.

A number of other projects had also been defined during the executive team meetings. These were divided into two categories—problems to be solved and systems to be improved. Two types of task teams were formed to deal with the two categories. *Problem-solving teams* worked on solutions to the specific problems. *Quality improvement teams* worked to improve the operation of entire systems—production, financial, hiring, and so on. Most of these groups were cross-functional, including representatives from any part of the organization management thought could contribute.

When the assignment had been completed, the task team was disbanded. Each team had a leader whose main responsibility was to keep the group focused. This person could be either an hourly employee or a manager. Group leadership rotated periodically so that everyone might get a chance. Each group also had a champion, usually the manager who initially came up with the project. The champion's main role was to support the group's recommendations when they reached the executive team. The executive team made all the final decisions based on long-range corporate objectives.

Eventually, any employee at any level could suggest problems or redesign projects to be addressed. Such suggestions were taken under consideration by management, prioritized, and assigned to a team. If it was decided that a suggestion should not be assigned, someone explained the reason why to the submitter.

Some such suggestions came from *involvement groups.* All employees were required to belong to one of 35 involvement groups that spanned the organization. These groups met monthly and also rotated leadership. The purpose of meetings was to reinforce teammate commitment to the TQM principles as well as commitment to the corporate values defined by the executive team. In addition, involvement group meetings were communications forums during which teammates shared information and discussed the progress of ongoing projects as well as the results of those completed.

In terms of the amount of paper generated by the Ames TQM process, an agenda for each team meeting was published. Minutes of the meetings were printed and distributed to team members. One copy from each team for each meeting was added to a central file in the quality office. Periodic surveys were also conducted. These included surveys on:

- Management practices.
- Team satisfaction.

- Cultural change.
- Success criteria measurement.

The results of these surveys were made available to all employees.

In terms of communicating results of the effort as a means of integrating and making sure everyone knew what was going on, several vehicles were used. One, of course, was the involvement group meetings. In terms of paper, there was the Dear Teammate letter that each employee received periodically from the CEO offering the view from the top. Also, there was the "Echo" newspaper which appeared quarterly and discussed projects in progress as well as successes. Recognition meetings were held during which teams and individuals were congratulated and rewarded for their efforts. And, finally, there were thank you type social events.

Few external consultants were used during the deployment phase. Their main purpose, when brought in, was to complement in-house staff for training purposes. Process-related questions could be directed to Xerox. The two major expenses, therefore, were the time spent by employees in training and team meetings and the recognition fund used to finance rewards for outstanding performance. Gifts, certificates, plaques, and monetary rewards were given to both individuals and teams. The executive team, in conjunction with the Total Quality Office, decided whom recipients should be.

The Total Quality Office at Ames Rubber currently is made up of seven staff members. Three of these are dedicated to quality assurance. Two are responsible for the quality documentation/ISO9000. One is responsible for quality education and training. This last person is in charge of coordinating the total quality training effort.

In terms of tangible improvement produced by the Ames Rubber Corporation process, the level of customer satisfaction increased; the number of repeat orders grew; there were major reductions in the manufacturing costs as well as in the price of products. These reductions resulted, at least partially, from a major reduction in the number of defects per million parts.

The Ames Rubber effort has obviously been successful. It has also, obviously, been top-down and management driven with managers supporting and controlling every aspect of the effort. The Ames Rubber effort won the Malcolm Baldrige Award. Now we are going to talk about a second organization that mounted a successful quality improvement effort, one in which the facilitator style of leadership was used. This is the Louisiana Mill of International Paper. The model instituted at this mill was not top-down but, if we must call it something, all over at once.

International Paper's Louisiana Mill

International Paper began its quality process like most other major corporations. The CEO made a speech. A quality improvement department was created. The quality department picked a highly visible, expensive consulting firm, Philip Crosby, to come in and do the initial, top-down training. A cross-section of middle-level managers attended a three-day seminar. They were then supposed to return to their units and put the process into place.

None, of course, succeeded. Many just faked it. Those who actually tried to practice what they had learned were frustrated. Owing to the brevity and to the classroom nature of their training, they had been provided with only fragments of the necessary whole.

The exceptions to this failure scenario were some of the mavericks who decided to ignore the corporate model and do it their own way. Bob Goins, manager of the Louisiana Mill, was one of those mavericks. His large, primary mill produced both pulp and paper. It employed approximately 1200 people. Goins found two internal consultants, not part of the quality department, who know something about team building. He brought them in and talked with them about how to produce quick results without spending a large amount of money up front.

The first step in the chosen approach was for Bob and the two consultants to meet in separate sessions with top-level, middle-level, and lower-level management and with the union local officials. These sessions lasted approximately one hour each. During these sessions Bob voiced his support for the effort. The consultants then briefly introduced the team building model, talked about its systemic origins, and talked about the amount of time workers would have to be freed up for team meetings.

The next step was for the head of Human Resources and the consultants to break the operation down into functions to be represented by teams. Function usually followed department, but there were some important exceptions. Two were maintenance and secretaries. Both were given their own teams to deal with problems specific to their function, rather than being assigned to the function they served, say the wood yard (maintenance) or the accounting department (secretaries). Not everyone could be on a team owing to the size of the organization and the continuous nature of the manufacturing process involved. Managers, therefore, were asked to assign members whom they thought would contribute meaningfully. These would eventually be rotated so that everyone who wanted to received a chance to participate directly.

There was no up-front training in process skills for team members or facilitators. It was possible, therefore, to bring the first team up soon after the functional breakdown was completed. One of the corporate consultants brought it up. A mill employee from the training department was assigned to work with the consultants. This person was designated head facilitator. He and the future team facilitator, who was from another function, sat in on the first meetings of the Finishing and Shipping team.

The first teams were composed solely of hourly workers. There was no designated team leader. The facilitator was also an hourly person. To begin the process, the consultant explained briefly the purpose of the effort to the members of each new team; the fact that it had the mill manager's support, though he, at this point, had turned control over to the work force; and the fact that if employees did not eventually find results beneficial to themselves as well to as to the company, they could drop off the team. The only thing being asked was that they give the process a chance. The consultant also introduced the team ground rules, which had previously been agreed to by all of management.

Next, the consultant led the team through a search conference exercise during which members identified a wide range of problems they were interested in working on. They then broke these problems down into three categories and categorized them. Team members were encouraged to put the ones which could be completed the quickest at the top of their list so that the team could produce quick results, proving to themselves that they had truly been empowered and could make changes; proving to top-level management that the team approach to improvement was not a waste of time and was not opening the doors to anarchy. By either the end of the third meeting or the beginning of the fourth, each team was working on its first project.

The consultant brought up the first two teams, then had the head facilitator bring up the third and fourth while sitting in support. After the fourth, the head facilitator was capable of bringing up teams on his own with the consultant on call to answer questions. Eventually, two more people were trained by the consultant to bring up teams. The person bringing up a team also ran the first several problem-solving sessions, which began in the third or fourth week, with the future team facilitator observing. Eventually, the team facilitator took over with the person who brought the team up in support.

Managerial teams were formed as well. There were of two types. One included all the managers from one function—foremen, supervisors, and superintendents. The other included all the managers from one level, say supervisors, across the mill. This latter type dealt with issues indigenous to its level. For example, exactly how much authority did the foremen have? Did the amount of authority vary from function to function? Should it vary? The lead team, composed of the mill manager and his direct reports, was responsible for heading the long-range planning effort, for developing strategic objectives. The other side of the coin was that the issues raised by the hourly and managerial teams were taken into account by the lead team in their planning effort.

The lead team was also responsible for stroking process participants. It listened to and complemented presentations on projects. Finally, the lead team was responsible for dealing with managers who were unwilling to obey the ground rules and who would not listen to the team facilitators or the head facilitator.

The hourly teams were called *problem-solving teams.* The managerial teams, with the exception of the *lead team,* were called *design teams.* A third type of team, the task force, could be formed by any manager around any project and could use anyone the manager thought necessary to the effort. Managers, however, were not allowed to jump the process and take over team projects they thought interesting. Also, they were not allowed to build a task force around the same project a team was working on. Task forces were dissolved once the involved project was completed.

The Louisiana Mill approach to quality improvement was relatively paper free. After the original project list was complete, minutes were taken for each team meeting listing in outline form progress on current projects, new action steps, and new projects picked. Team members were allowed to distribute their minutes to anyone they wished. One copy was also kept in the master file in the quality office. No surveys were taken.

The quality improvement department consisted of the head facilitator and a part-time secretary. In terms of communicating process results, a sheet of flipchart paper with each team's name on it—Paper Machine Number 1, Secretaries, Finance, Warehouse, Wood Yard, Pulp Machine Number 1, and so on—including a list of ongoing and completed projects, was hung in the entryway to the mill. Almost everyone passed through the entryway daily and could stop to look the sheets over.

Specific projects were discussed in more detail in the monthly mill newspaper. Mr. Goins began holding open meetings at which the progress of the planning exercise was discussed and for which employees could submit their own agenda items, anonymously if desired, through the head facilitator. Also, videotapes were made recording some of the more creative improvements.

These, however, were not so much for recognition as for distribution to other mills addressing the same problems.

The only cost, technically, was the travel time of the two in-house consultants and the time spent in meetings. Because the teams started producing results almost as soon as they were formed, however, this cost was very soon covered.

One issue which came up almost immediately was job security. The mill manager issued a statement to the effect that it would become his number one priority. While he could not ensure that employees would not be moved around in the mill, if the quality improvement process produced positive bottom line results he would do his best not to lay anyone off. And he kept his word. In fact, while other mills were laying employees off, he was adding staff made possible by the growth in profits.

In terms of rewards other than job security, the only one offered initially was the fact that employees gained real ownership of their part of the operation. This seemed to suffice. After a year or so the mill manager and his reports began having lunch with teams. Also, communication between managers and workers increased and grew a lot less formal.

There were never, however, any monetary rewards for completed projects or for the best project. This tactic was considered not necessary at best and divisive at worst because it created competition where cooperation was the key. Eventually, there was talk of some form of profit or gain sharing, but this had to be approved at the corporate level.

Like the Ames Rubber Corporation project, the Louisiana Mill project was a success. Within a year 19 hourly problem-solving and 23 managerial design teams were in place. Task forces came and went and eventually adopted the same ground rules used by the quality process teams. Within 1½ years the teams had completed over 250 projects and the mill had set records in 20 major categories.

Exploring the Differences

One major difference between these two models is obvious. The Ames Rubber Corporation process, as we have said, was implemented top down. It was management driven and controlled. The initial focus was on training, with top-level managers being trained first, then being made responsible for training reports, and so on down through chain of command.

A major benefit of this approach is that it sent a strong top-down message. At the same time, it introduced the concept of quality improvement without forcing major and possibly traumatic shifts in the organization culture. It did so by providing what we call partial, rather than full, empowerment.

The Ames Rubber model also obviously improved organization-wide communication. It heightened understanding of the organization's mission and objectives. It developed a common language and common mind-set. It forced managers to become more responsive to their reports. It got all employees involved, to some degree, in making improvements.

Finally, the Ames model locked managers and hourly workers in. With such a strong mandate coming down from the very top, with so many hours of training under their belts, it was hard for bosses, and closet bosses, to say, "Hey, wait a minute. I do not really like this. I like the old way better. I like making all the decisions. It is more efficient, more comfortable doing it the old way." With the CEO's presence and determination looming over the process, bosses were forced to at least pretend that they accept the change, to play along, or to leave the company.

In the International Paper model this pressure does not exist to the same degree. This second model is, as we have said, all over at once. It is as much bottom-up as it is top-down. Its primary objective is to fully empower employees on all levels. The CEO helps introduce the process and voices his or her support for it, but from that point on he or she steps back and lets it happen, functioning as a facilitator. In this model the initial focus is not on training. Rather, it is on gaining employee commitment by granting process ownership on all levels.

The Louisiana Mill effort assumed that all employees understood the need for improved quality and were interested in contributing. The Louisiana Mill effort, therefore, took a lot more for granted. At least partially because of this abbreviated approach, the IP effort took a lot less time to implement than the Ames process—approximately one year for 1200 employees as opposed to four years for 600. It also cost a lot less in terms of prep time, and began producing tangible bottom line results more rapidly—approximately two months as opposed to three years.

The Louisiana Mill model did not lock employees in up-front. Its philosophy was that employee commitment would be more real if employees were allowed to make up their own minds. The process had to prove itself. Each employee had to be convinced that it would benefit him or her individually. If that conviction was not generated, if employees were not convinced that they were going to gain personally in some way, no amount of training, no amount of top-level support was going to keep it alive, at least not in the long run.

As was said earlier, all employees were required, at least, to give the process a chance; but after the first four or five weeks of participation, if they did not see any value to it they could drop out, they could call for a change in the model, or, if enough disbelievers banded together, they could just plain shut it down. First line- and middle-level managers were, of course, the greatest hurdle. But the game was the same for them. The process had to prove itself beneficial, or they did not have to play.

Another difference between the two models, in terms of full vs. partial empowerment, was the way in which team projects were generated. At Ames Rubber, at least initially, all projects were identified and assigned by the CEO's group. Next, managers were allowed to define projects and to build teams around them once they had been approved by the CEO's group. Eventually, all employees were allowed to suggest new projects, but these, again, had to be reviewed, accepted, prioritized, and assigned a champion by the executive team.

At International Paper, while management could continue to build a task force around any project it wanted, and could put anyone on that task force it considered necessary, the bulk of the improvements were generated by hourly and managerial quality teams that built their own project lists, defined their own priorities, and controlled their own projects, within the bounds of the ground rules.

This is a critical difference between the two approaches. It is a difference similar to that between giving employees real ownership of the change process, and giving them shares of stock in it, but making sure that they never gain a controlling interest.

One rationale offered for this latter attitude, as we have said, is that top-level management must maintain control in order to integrate the effort effectively. Top-level management has defined the organization's long-range plan and objectives at the onset of the process. It best understands them and, therefore, best understands what should be worked on.

The rebuttal is that in the Louisiana Mill model long-range planning is an ongoing part of the process. All teams contribute, in some degree, to the definition and fine-tuning of objectives, so all employees are familiar with them. In terms of their own area of expertise, their own 25 square feet, because of their greater familiarity, they know best what improvements are necessary and how to prioritize them in order to achieve the organization's overall objectives.

One must also ask whether in the Ames model employees truly buy in, or whether they do so because it would obviously be suicidal not to. We are not insinuating that they do not buy in, but the question is still open. With the all over at once model there can be no question about commitment. If employees do not buy in, they have the power to shut the process down.

With managers especially there is initially a lot of skepticism. They have seen it all before. They, too, are forced to give it a chance, but that is all. Some become convinced that it is worth a try when the hourly teams begin making improvements almost immediately, when the number of mini-crises they have to deal with daily decreases, when they get more time to work on the things they think important. Others become convinced when the process gives them their own teams and empowers them to work on projects they never thought possible. And some, of course, are never convinced.

In the Louisiana model this latter group was not allowed to hinder the process, to keep hourly workers from attending meetings, but no pressure was put on them to join in. Some workers, some managers, usually old-timers close to retirement, could never accept the new challenge and clung to the old way. But as the process moved forward and the results rolled in, their numbers tended to become inconsequential and their attitude one of grudging tolerance rather than resistance.

And According to our Systems Criteria

In terms of our systems criteria, then, there are again a lot of similarities, but also serious differences. Both the Ames Rubber Corporation and International Paper mill projects included the five necessary phases. Both included a process success measurement vehicle and a reward system. However, the approach to each of these elements was different in terms of the characteristics necessary to the systems approach—participative, holistic, ongoing, and providing continual feedback. Although these differences were mainly a matter of emphasis and degree, they affected outcomes and should be examined.

The familiarization phase at Ames Rubber was much more comprehensive. A great deal of time was spent up front. Involvement groups were put into place to keep it alive. The Ames Rubber familiarization process was obviously participative in the long run, organization-wide, ongoing, and provided continual feedback. It gave employees a degree of control and fostered cooperation, but it also consumed a great deal of time. The company obviously considered this expenditure important.

The International Paper mill model reduced the up-front piece to two weeks. Its philosophy was to familiarize mainly through demonstration rather than up-front instruction. The ongoing effort was also far less formal. The belief was that as communication systems improved, answers to questions would be informally available on any level. Representatives of the facilitator network were always available to groups with questions. The mill manager held a monthly meeting where a progress report was given and anyone could ask anything they wanted. A newsletter was also published monthly and team projects were posted on the wall of fame.

The International Paper mill familiarization phase, therefore, was also participative, organization wide, ongoing, and provided feedback, but in a much less structured manner. The International Paper mill model gave more control and responsibility to the individual employee; but, at the same time, by not mandating participation this model might have lost the opportunity to convince some employees who had doubts and, therefore, avoided change-related activities. The International Paper mill model of the familiarization phase, therefore, had the potential of generating a smaller degree of participation.

In terms of long-range planning, the Ames model was not interactive, and, therefore, not systemic. Planning was done at the top, with no input from other levels, the results being presented to the rest of the work force. Quality improvement projects did result from the planning effort, but again, most employees had no input. The process, therefore, was not participative. It was organization-wide and integrated because a cross-section of managers were involved. It was not ongoing. The planning effort took place at the beginning of the process. Quality projects were shaped by the long-term objects defined, but, in turn, did not help shape, or reshape, these objectives. Because the Ames planning effort was not ongoing, there was little chance for feedback.

The International Paper planning effort began at the same time as team building and was linked to it so that team projects helped shape the plans just as the plan provided a frame of reference for the projects. Because it was linked to team efforts and because team members were fully empowered, the planning effort was participative. Because it was linked to the quality team and task force network, it was organization-wide and integrated. Because it continued to evolve as team projects evolved, it was ongoing. Finally, for these same reasons, there was constant feedback flowing in all directions.

In terms of team building and team-related activities, the Ames Rubber model was semiparticipative at best. Only task forces existed, though some were longer-term and worked to improve entire organization processes. Managers usually led the task forces. Each task force had a champion, who was also a manager. Projects were assigned by upper-level management. The majority of employees involved, therefore, were not empowered in the systems sense.

The network of task forces at Ames was not structured to be organization-wide. The company's approach was analytical. It focused on specific pieces of the puzzle. Integration occurred only at the very top of the management hierarchy, again failing to empower the majority of employees. Individual task forces, by definition, were not ongoing, making integration of the network even more difficult. Feedback existed, but, again, it was the up-and-down kind, rather than all over at once.

Whereas the task forces in the International Paper model, just like those in the Ames model, were not fully participative, the quality improvement teams were. This was mainly because workers and managers were separated, and because the teams picked their own projects. The quality team network at International Paper was organization-wide, representing every function, and integrated by the ground rules and the facilitators. It was ongoing; teams continued to develop new projects. Feedback occurred in all directions owing to linkage with the planning effort and with the ground rules.

In terms of training, concerning both group process and statistical measurement skills, it was done mostly up-front at Ames. It was done to the employees. They had no input, so that it was not participative. Future, job-related training needs were defined participatively by task force members. The training at all stages was organization-wide and integrated, as well as ongoing. The vehicle of surveys also allowed feedback concerning effectiveness and future training needs.

At International Paper the training was incorporated more so into the other phases so that it would be more meaningful and so that participants would have more say in shaping it. In terms of group process skills, the apprenticeship model was used. Facilitators, especially, learned on the job. Facilitators also identified their own future training needs. They frequently designed and carried out the involved training themselves.

Statistical skills were taught to facilitators in the classroom at International Paper, but quality improvement team members had to identify a need for them before they were introduced. Quality team members, like task force members at Ames, also identified a large number of job-related training needs. Training at International Paper, therefore, was largely participative in all stages. It was also, obviously, organization-wide, integrated, and ongoing, with continuous feedback coming from everyone in the network.

The Lessons We Want to Learn

When we are trying to improve organization performance, although more than one useful model might exist, those which are the most systemic tend to be the most effective, produce the quickest results, and provide the greatest benefits from a developmental perspective to key stakeholder groups.

The partial empowerment, management-driven model is the one currently in vogue. It is, for example, the model currently winning the Malcolm Baldrige Award. The partial empowerment model, as we have said, is a great improvement over the traditional autocratic model and has shown its value. It has quite a few systemic characteristics.

The total empowerment model, however, is more systemic and developmental in nature and is probably the model of the future, of new world organizations, and of the development ethic driven society we are striving for.

Some countries are closer to having it in place than we are. The Scandinavians, as we have said, practice full empowerment in many instances. Australia is another example. Fred Emery, one of the true gurus, and his students are fairly well along. So are the Japanese, who have gone through the same stages we are encountering. Their first quality circles, some 20 years ago, were led by managers and were assigned projects by top-level management. Then the Japanese began to allow circle participants to pick at least some of their own projects. Finally, a growing number of Japanese firms have begun forming circles composed entirely of hourly workers, facilitated by hourly workers, and have begun encouraging these groups to pick their own projects.

The same progression is occurring in the United States. We are behind, but it is happening. Leaders in U.S. companies are slowly adopting a new role in the workplace. They are gradually becoming equal members in the work team instead of heading it. Their new role is gradually becoming that of facilitating the release of individual and group potential. It is gradually becoming that of helping coordinate and integrate individual and unit efforts and of helping team members understand the big picture.

The personal skills of leaders and facilitators in this new scenario are becoming more closely integrated with and are as important as, but, in terms of status, no more important than the skills of other team members—

technicians, marketing experts, researchers, accountants, engineers, maintenance personnel, and trainers. This change in perspective will totally flatten the management hierarchy, but in a realistic way. Instead of firing leaders because their traditional role as boss or manager is no longer relevant, instead of foolishly dumping all that expertise out onto the street, we will realize that the problem is not one of leaders losing their value, but rather one of an outdated paradigm, of our outdated concept of hierarchical management, and of, "Somebody has to be in charge!"

We will take the time to redesign our organizations and to redefine the responsibilities of our current leaders so that their expertise can be again used effectively. We will meld them with the rest of the work team, rather than having them watch over it from their elevated position in the organization chart. In the resultant new world model there will still be a hierarchy of sorts, but it will involve levels of integration rather than of leadership, and will be built around policy/planning teams or boards like those in Russell Ackoff's circular organization.

Now it is time to get into the details of the systems approach to quality improvement. We shall do so by presenting a comprehensive case study of the International Paper Company effort, leaving nothing out, discussing the failures as well as the successes, the mistakes as well as the good ideas, the tactics of those opposed to as well as of those who supported the involved change. The model worked beautifully, as we saw in this chapter, but, in the end, in terms of the big picture, the effort failed. It did so because a key ingredient was missing. We shall eventually understand why it failed, despite its success, and what that missing key ingredient was.

Topics for Discussion

1. What role did top-level management play in the two models?
2. What role did teams play in the two models?
3. Compare the way efforts were integrated in the two models.
4. Why is the Ames model called partial empowerment and the International Paper model called full empowerment?
5. Compare the overall expense of implementing the two models.
6. Why does the Baldrige model continue to be more popular?

7 Birth of the Systems Model at International Paper

After Reading This Chapter You Should Know

- How International Paper (IP) Company started with the traditional, top down, training-driven model.
- How and why input from the Organization Planning and Development (OP&D) department was rebuffed.
- How the corporate model began to falter almost immediately.
- The difference between operational and improvement-related problems.
- The five characteristics of systemic team building efforts.
- What the major problem with the management system at the Louisiana Mill was.
- How the first hourly team was started and the process-related problems that arose.

The Right Mix

During 1985, one of IP's 16 primary mills, the Louisiana Pulp and Paper Mill, accounted for approximately 20 percent of the corporation's total profits. In so doing, it generated twice as much profit as its nearest competitor. The mill repeated this performance in 1986 and again in 1987.

The mill manager, Bob Goins, was one of the most experienced in the company. During his ten years at the Louisiana Mill he had put together a highly regarded team of direct reports. These two facts alone, however, were not enough to account for the sudden and dramatic increase in profits. The mill had invested in new technology. The wood yard, for example, had been reorganized. A new boiler had been built. A new computerized control system had been installed, but again, this investment was not considered the only producer of the upturn. In fact, a number of critical problems remained to be worked out with the new technology before its effects on profits could be truly felt. At the same time, other mills in the system, some with more capacity, had also been reorganizing and computerizing, but had not enjoyed the same improvement in their bottom line. What, then, was going on at Louisiana?

On June 16, 1987, a strike began at IP's largest manufacturing facility, the Androscoggin pulp and paper mill in Jay, ME. Despite appeals from union officials not to fill the 1200 hourly openings, applicants, attracted by the relatively high wage, quickly did so. A majority of the replacement workers had no experience in paper making. Supervisors and staff people left their offices to conduct on-the-job training and to keep the machines running. Maintenance personnel were sent in from all over the company to make much needed improvements and to help repair the damage done by strikers when they left. Workdays of 12 to 18 hours were not uncommon.

The strike dragged on for 16 months. Violence occurred. Car windows were smashed. Nails were strewn on roads to puncture tires at shift change. Fights broke out. Shots were fired into the homes of both replacement workers and supervisors. Most of the replacement workers lived 50 to 100 miles from the mill. Each day they had to drive through a gantlet of jeering, cursing pickets to get to the mill gate. When they headed home at shift change the pickets were again waiting. Moral was low, exhaustion widespread. Several serious industrial accidents occurred.

Less than two years later, this same group of relatively inexperienced hourly workers and old-time supervisors set an all-time production record. Two months after setting it, they broke it. Three months later, they broke it again. Quite a bit of money had been spent to improve the mill's technical systems, more than was spent on any other production facility during this period. Most employees, however, agreed that a major factor in this success story was the change that had occurred in the mill management culture. It was the same change that Bob Goins credited for precipitating the exceptional performance of the Louisiana Mill. Its source in both situations was the systems approach to quality improvement discussed in Chapter 3 of Part I.

The purpose of Chapter 5, Part II, then, is to show how this systems approach was successfully implemented in two extremely different real-life situations. Describing events in their chronological sequence will provide our framework for this section. Emphasis, however, will be on developing a feel for what happened. What we are talking about is a cultural change effort, and, as has been pointed out in the literature many times before, cultural change is as much or more so a matter of emotion than it is of intellect. The feel for such a process that people get in their gut is eventually as important as or more important than the understanding they achieve in their brains.

Two events critical to our history occurred at IP corporate during 1983. The first was the kicking off of a quality improvement process (QIP). IP followed the traditional route. Senior executives visited companies that had mounted successful processes, including General Electric, Ford, and Westinghouse. They talked to disciples of Deming, Crosby, and Juran. Then, after deciding on the Crosby approach, they generated corporate goals, a quality policy, and management principles that complemented the goals and policy.

The leader selected for the effort was a bright, young star with an engineering background. The first task he and his staff undertook was to organize seminars. Approximately 40,000 employees were introduced to the new corporate goals, quality policy, and management principles. Approximately 13,000 employees were familiarized with team building skills and statistical quality control techniques. Finally, a head facilitator chosen from each facility in the company received several additional days of intense training.

The new vice president of QIP reported directly to the corporate CEO, John Georges. His staff consisted of four specialists. One was responsible for developing measurement tools. A second was in charge of ongoing development of the corporate approach. The other two were in charge of facilitating and guiding the efforts of the over 100 manufacturing facilities and support units scattered across the North American continent.

The second, initially unrelated event, was that Ron Cowin, Director of OP&D of the Human Relations Division, reorganized his staff, bringing in three new people. One was Doug Ferguson, who had been with IP for ten years. Doug had started as a cost analyst. He worked as the Manager of Finance at the Ticonderoga Mill, a Senior Analyst with Strategic Planning for Paper Board and Packaging, and a Senior Analyst for New Products Development, before coming to OP&D.

Joyce Avedisian's background was in organizational behavior. She had a Ph.D from Brandeis University and had worked previously with the Arthur D. Little Company and as an organizational development consultant with AT&T.

The third person, Tom Roberts, had never held a job in the private sector before. He had earned a Ph.D from the Wharton School in the systems approach to management sciences and had consulted in both the private and public sectors for a number of years. Most recently he had taught management science at LaSalle University in Philadelphia.

Ron's desire in bringing these three people together was to make possible a systemic approach to his department's assignment. Doug possessed experience in strategic planning. Joyce's training was in individual and group interaction; and Tom had a strong academic and consulting background in organization design. Between them they possessed expertise in the definition of long and short-term organization objectives (the essence of the functional approach); the facilitation of employee participation (the essence of the process approach), and the improvement of unit/department/division linkages and interactions (the essence of the structural approach).

First Try

After attending the corporate quality training seminar, members of the OP&D Department felt strongly that they could contribute to the quality improvement effort. The corporate approach, at this point, focused mainly on defining *what* had to change and *why.* OP&D now possessed the expertise to help define in a comprehensive manner *how* to actually go about implementing the desired changes. Members of the department sat down and developed a systemic model based on the work done in earlier projects by systems practitioners. A meeting was scheduled with the staff of the Quality Improvement Department and the model was presented. Having just spent several millions dollars on the Crosby approach, however, the Quality Department's response was not enthusiastic.

OP&D decided that it should test the systems model somewhere in the company on its own in order to prove its value. Ron thought that OP&D's first target should be a headquarters unit, but this plan did not go well either. The realization was that what OP&D was proposing looked too much like a bid for increased exposure and, therefore, was going to be contested or ignored by too many of those whose support was necessary.

The next step was to seek a test sight at a mill. Because of Doug's recent employment at the Ticonderoga Mill and his reputation there he was able to set up a meeting with the manager. Mill personnel had attended the quality training seminars. Upon their return they had held the necessary sessions and distributed the required materials. After that, however, very little had happened.

Several things quickly became obvious during this initial visit. First, what the Ticonderoga Mill had put together was, in essence, an aggregate of loosely connected lower-level teams whose primary role was to implement quality improvement suggestions passed down by upper-level managers. Second, the training of those responsible for facilitating this process was minimal at best and frequently nonexistent. Third, those who had designed IP's approach to quality improvement had made a familiar mistake. They had given responsibility for putting the process into place at the hourly level to middle management, to those men and women who would be most threatened if the process succeeded and lower-level employees started playing a more active role in decision making.

Doug and Tom proposed to the manager of the Ticonderoga Mill that as an initial step a more comprehensive team network be created. The operation would be broken down into major functions—power, manufacturing, maintenance, finishing, and shipping. These major functions would then be broken down by shifts at the hourly level. Each shift would be represented by a 10 to 12 person problem-solving team. Team membership would rotate so that everyone on the shift eventually participated. In order to win the support of both team members and those they represented, instead of implementing improvements mandated by management, Doug and Tom suggested that teams be charged with identifying and implementing improvements they themselves thought important.

In terms of the middle management issue, rather than butting heads with these people and expending the time and energy necessary to win their cooperation before attempting to move the process on down into the hourly ranks, it was suggested that the effort simply circumvent them. Once the teams started producing positive results, it would be much more difficult for middle managers to balk and much easier for them to see the value of the process.

The mill manager and head of operations both seemed willing to give the approach a chance. The one person with serious reservations was the head of the mill quality improvement program who reported to the Vice President of Quality at corporate headquarters as well as to the mill manager. His major concern was control. He wanted Tom and Doug to report to and be responsible to him, rather than to the mill manager.

This was a serious matter, for as an engineer he had little training in management systems modification beyond the two-day seminar and some reading. He suggested strongly that most of the things discussed were already in place or were not necessary. It was definitely a turf issue. Doug and Tom were not even members of the Corporate QIP Department staff. Yet they were about to usurp part of his responsibility using as their rationalization the claim that they wanted to complement rather than replace the corporate approach. Corporate was coming top down. What OP&D offered was coming bottom up. Corporate understood the need for teams, but could not get the necessary network into place. OP&D was there to help with that part of the process.

The mill manager eventually decided to proceed, owing mainly to his relationship with Doug. In the middle of negotiations, however, it was learned that he was being transfered from Ticonderoga to another position in the company, and the project was put on hold.

Opportunity Knocks

OP&D's second opportunity came through the joint efforts of Ron and Doug. The manager of the Louisiana Mill, one of IP's largest primary mills that produced both pulp and paper products, had begun a project to develop a state-of-the-art finishing, storing, and shipping division (PS&D). To identify the changes necessary, he had decided to form a task force of internal consultants with the necessary technical expertise. When Ron, who had held a series of conversations with him, suggested that Doug be made part of the task force, the mill manager consented.

The initial meetings were held in late December and January of 1985. When Doug eventually suggested that the employees themselves be asked to suggest improvements, some of the internal technical consultants objected strongly. The mill manager, however, decided that the two approaches—expert driven and employee driven—should be mounted simultaneously.

The task force's agenda included the study and redesign of systems such as:

1. The total entry order system from conception of the order.
2. Through planning and scheduling.
3. Equipment systems.
4. Performance measures for both the system and equipment.
5. The quality of services offered.
6. Organization structure.

7. Engineering standards.
8. Training.

Doug's assignment was to review, with input from PS&D staff, all possible organization structures suitable to PS&D, to define the strengths and weaknesses of each, and to review the current PS&D communications system and suggest ways to improve it. This assignment fit well with OP&D's objectives. It gave the department access to PS&D employees. It provided as a twin focus the division's structure and its communication system, two critical component of any serious organization change effort.

Back in New York the OP&D staff began putting the model to be used at the Louisiana Mill together. All it actually had at this point was the sketchy straw man developed at Ticonderoga and the systems concepts agreed upon during earlier staff meetings. The starting point decided on for this exercise was that in all businesses two types of problem-solving go on. The first was labeled operational. This involved everyday issues that affected the mill's ability to survive and that required immediate attention. The focus here was on getting the daily requirement of sheets, rolls, and so on, out the door. Operation problems were usually generated by the production process itself—sudden machine or computer stoppages, the loss of orders, sick or negligent employees. Corporations locked into crisis management are usually those that could not get beyond operational problem solving.

The second type of problem was labeled improvement-related. This included efforts by employees on all levels to improve the operation based on their experience and ideas. Improvement-related problems did not usually require immediate attention. Examples included discovering the best way to shorten an operating procedure, developing a more effective training program, improving an interface between shifts or between departments.

A strong love-hate relationship existed between *operational* and *improvement-related* problem-solving efforts. On the love side, while the former were short-term in perspective and concentrated on keeping things going, the latter came from a longer-term perspective and focused, at least partially, on decreasing the number of daily operational problems that had to be addressed. Obviously, everyone was in favor of making as many improvement-related changes as possible.

On the hate side was the fact that these two types of activity competed for time and resources, operational problem-solving efforts necessarily winning out. All supervisors wanted to work on improvement-related issues. First, however, they had to deal with the mechanical problem causing a slowdown, then with an employee's drinking problem, then with a critical shortage of replacement parts, then with an unhappy customer, and so on.

Once one understood the hate side of the relationship, it was not difficult to figure out why employees gave up suggesting improvements. It was also easy to figure out why so many corporations were beginning to suspect that long-term planning was mainly a waste of time.

The Model

With the above in mind, the OP&D staff shaped a model for the Louisiana Mill that had *five key characteristics.* The *first* was that it segregated, at least initially, improvement-related problem-solving efforts from operational efforts. This was necessary if the former were to receive adequate attention. A network of teams would be formed. These teams would meet for one hour a week, during which time they would focus solely on systemic improvements.

The *second* key characteristic was that, as suggested at Ticonderoga, the team building effort would begin in the hourly ranks. As quickly as possible, hourly employees would be given the chance to make reasonable changes in their areas of expertise.

A *third* key characteristic of the model developed for Louisiana was that, while team building would begin bottom-up, top-level management needed also to play an immediate role in the process. If possible, a planning expertise should be started. The lead team should put together an idealized version of the PS&D Division, thus providing a framework into which to fit the improvements coming out of hourly teams. Such participation would, in all probability, be a learning experience for top-level management. It would also subtly demonstrate the need to expand the project mill-wide. Finally, it would make top-level managers more accessible to the lower-level problem-solving teams as the two efforts began to mesh.

The *fourth* characteristic concerned the role of the OP&D consultants. This role would be to make sure all employees developed an accurate understanding of the approach; to help gain the support of upper-level management, specifically the mill manager; to bring up the teams; but most importantly to train and organize a network of facilitators, who would be responsible for maintaining the teams once the consultants left. A head facilitator would be appointed by the mill manager and would work with the consultants to help plan and guide the overall effort. An obvious candidate for this roll was the mill head of Human Resources.

The *fifth* characteristic was that the team network had to be effectively integrated. A way had to be developed to deal with problems affecting more than one team, to avoid repetition of effort, and to keep the teams informed of what others were doing so that they might maintain a division-wide and mill-wide perspective.

The last thing that the OP&D staff put together was a draft of the process ground rules. Thus armed, they headed for the Louisiana Mill.

Getting the Ball Rolling

After gaining the mill manager's acceptance of their plan, Doug and Tom met with the head of the mill Human Resources Department to break the PS&D Division down by function. The breakdown decided upon was customer service, roll finishing machine A, roll finishing machine B, sheet finishing, and maintenance. Roll finishing A and B each included 50 to 60 employees, as did sheet finishing. Maintenance included 30 employees. Teams of 10 to 12 would represent these 4 functions.

The major strength of defining the teams strictly according to function rather than by shift would obviously be commonality of interest. Also, each function would have the chance to get its own sandbox in order before addressing its relationship with other functions. In terms of improving communication, a great number of the problems and misunderstandings which existed between shifts could be dealt with immediately when the teams included representatives from all four. The challenge of this approach, of course, would be getting team members together when they worked at three different times during the day and night.

The mill manager supported the decision to form the teams around functions. If this arrangement did not work, it could be changed, but he was fairly sure that it would work. Team members would not be volunteers, at least not initially. They would be appointed by the foremen and supervisors and would be required to attend meetings, even during off-hours, until they understood what the process had to offer. If eventually they felt they had no contribution to make or remained highly dissatisfied with the loss of one hour a week of free time, a substitute could be found. Those coming in during off-hours would be paid overtime for the time spent in meetings. Foremen and supervisors were instructed to choose the employees they thought capable of making a contribution in terms of both experience and attitude.

The question arose as to whether maintenance should have a team of its own or whether representatives should sit on and act as a resource to the other divisional teams. With input from the mill manager, the head of PS&D, and the head of maintenance, Doug and Tom eventually decided that it should have its own team and should deal initially with issues internal to its own operation. This arrangement would allow maintenance personnel to concentrate first on their own procedures and equipment needs. It would allow them also to design, from their own perspective, the desired relationship with the storeroom and the production units they served before getting other stakeholders involved.

At the same time, the ground rules guaranteed the rest of the teams access to members of the maintenance crew on an as-needed basis. This arrangement would prevent time from being wasted. Maintenance personnel would not have to sit and listen to members from, say, the Sheet Finishing team discuss issues in which they played no role.

The next point addressed was whether or not foremen or other supervisors should be made members of the hourly teams. Corporate policy was that they should. A major purpose of the team building effort was to improve relations between the hourly work force and management. Having the two sides join together in a problem-solving effort was a step in this direction.

Tom and Doug disagreed. The most important purpose of the team building process was to increase the commitment of the total work force to improve quality. This could be done on the hourly level by showing the workers that management respected and desired their input and ideas. The teams were a vehicle for encouraging them to speak up. If a foreman or supervisor was a team member from the start, however, it would be business as usual. The hourly workers would expect that person to take the lead and would be nervous about making suggestions he or she might disagree with or find threatening. At the same time, the foreman would be put into an awkward situation. It would be difficult to give an opinion without appearing to be attempting to dominate.

The ground rules ensured management's input. But that input should be *invited* by hourly team members, rather than forced upon them. They could eventually ask a popular foreman or supervisor to join the team, but the decision should be theirs.

Another point addressed by the mill manager before the process began was who the team facilitators should be. Mr. Goins felt strongly that young engineers should be used. The relationship between these people, frequently fresh out of school, and the more experienced mill hands and maintenance staff was not always good. Mr. Goins felt that the interaction forced by making them hourly team facilitators would lead to an improvement in communication. It also would develop a better understanding on both sides as to what the other had to offer.

Finally, Doug and Tom observed that, in order to truly succeed, the team building process needed to be implemented mill-wide. The interdependencies that were bound to crop up during problem-solving activities would make this necessary. Mr. Goins said he understood, but that the effort should precede as scheduled. It should start with PS&D to see how things went.

This Is What It's All About, Folks

The next step was to briefly introduce OP&D's approach to team building to all levels of management. Tom gave presentations to the mill manager's direct reports, to union representatives, and to superintendents and supervisors from PS&D. He began each presentation talking about industry's growing realization that employee involvement was critical not only to increased competitiveness, but frequently to survival. The old ways did not work anymore. There was too much to know in any modern operation for one person or a small group of people to have all the answers.

He then presented the bottoms-up model developed for the Louisiana Mill. He stressed the point that once the introductory sessions were completed, the OP&D consultants were going to focus on bringing up hourly level teams. These were the people the corporation had experienced the most difficulty in getting involved. They were the ones who had been left out of the up-front Crosby training, the ones most skeptical concerning the seriousness of the effort and its chances of producing lasting change. At the same time, they were the ones, ultimately, upon whom success rested because they actually produced the pulp and the paper.

Tom's presentation to the mill manager's direct reports was received in silence. No one said anything. No one asked questions. No one even looked at the speaker. Mr. Goins was present. His reputation as a mill manager was that of an autocrat. He was respected for his knowledge of the industry and was considered to be fair in the long run. But no one wanted to get onto his bad side. He was extremely demanding and rode employees hard, often in front of others, if he thought they weren't doing their jobs. At the same time, he fought for the people he believed in, often against superiors. He was considered the maverick of the mill managers. At least two of his direct reports did not possess the credentials, especially academic, deemed necessary for their positions and would not have gained them if Mr. Goins had not insisted.

Tom's presentation to the PS&D foremen, supervisors, and superintendents brought the same response. Again, Mr. Goins was there. Another reason for the silence was that most attendees were sitting in the presence of their immediate bosses. Tom talked about the current trend toward reducing middle management staffing levels to cut costs and increase the efficiency of operations. He said that the systems model did not accept the logic behind this move. Too much valuable expertise was lost. The effect on overall employee morale was too great. The long-term negative consequences of such a tactic far outweighed the short-term positive ones. Instead, a new, more productive role for middle managers should be found.

While a few members of the audience seemed to like what they were hearing and nodded approval, the majority stared blankly.

The presentation to the union representatives brought an entirely different response. Mr. Goins was not present at this session. A number of questions were asked. There were requests for elaboration. Finally, one representative said, "We've been asking for something like this for 20 years." Tom told the audience that the OP&D consultants would keep them fully informed as to what was happening and would be available to answer any questions.

One of the questions that did come up in all three of these presentations was, "How much of the employee's time will these meetings take?" Tom's answer was that following the start-up sessions each team would meet weekly for approximately 1 hour until it had progressed far enough to set its own schedule. Several of the managers said that they did not see how they could afford to release workers from their responsibilities for that long.

Maiden Voyage

The OP&D consultants had, by this time, decided that, because of his systems background, Tom would take the lead in the team building effort. He would immediately begin training Doug and Joyce in the necessary techniques. Hopefully, he would eventually be able to train several in-house people as well. Joyce would function as chief troubleshooter. Employees with serious doubts about the process would be directed to her. Doug would facilitate the lead team planning effort when it started.

In terms of their initial effort in PS&D, the consultants decided to start at the tail end of the operation and back into it so that improvement would immediately "go out the door" to customers. Based on this decision, the first team formed was Customer Service. This unit was responsible for scheduling paper production runs and for getting the finished product to customers. Because of the department's size (twelve hourly employees), it was decided to include them all on the team. Mr. Goins agreed to close the office for the start-up exercise, which was scheduled for the week of February 5, 1985.

When Tom and Doug arrived for the start-up the customer service staff was cheerful about getting time off from their normal duties, but suspicious. Tom did not take the time to explain the process as he had for management levels and union representatives because he did not want to get bogged down answering questions. Instead, he immediately began a Search Conference problem identification exercise that took roughly four hours. The end result (See Part 5, Table 7.1) was a list of 29 issues that were discussed briefly, then broken down according to three levels:

1. The issues that could be dealt with strictly by customer service staff.
2. Those that involved another mill function.
3. Those that were mill-wide.

Table 7.1 First Search Conference Session Notes for Customer Service

1. Trends in United States community that affect quality of life:
   - Exporting of jobs as industry moves out of country.
   - Bad exchange rate.
   - Immigrant labor taking jobs.
   - Growing national debt.
   - Growth of trucking industry, decline of rail use.
   - Rising personal taxes.
   - Good national governmental leadership.
   - Increased governmental involvement.
   - Decline in the quality of education.
   - Increase in drug, alcohol use owing to:
     - increasing stress.
     - poor judicial system.
     - peer pressure.
   - Decline in morality.
   - Young people want to help but lack leadership opportunities.
   - Greater competition overall.
   - Growing pressure, demands.
2. IP trends:
   - Diversification of IP into areas where we lack expertise.
   - Management wants more control over little people.
   - Management too rigid, lacks flexibility.

(continued)

- No back-up orders in system.
- Crisis management, little time spent planning ahead.
- Priorities undefined.
- Must rely too much on corporate to get job done.
- Lack of appropriate tools to get job done.
- Rules not followed by everyone.
- Poor communication systems overall:
  - – within mill.
  - – between mills.
  - – between mill and corporate.
  - – within customer service department.
- No coordinator at corporate.
- Lots of buck passing.
- Too many contacts at corporate.
- Information necessary to job often unavailable or old, useless.
- Bottlenecks in the information system.
- People not answering phones/radios when called, not returning calls.
- Data processing too far from customer service.
- Warehouse too small, layout.
- Every time more space becomes available, corporate fills it up.
- No one responsible for the warehouse, no supervisor, paper stored all over the place, no pattern.
- No delegation of authority.
- No set pattern, no system.
- Priorities change constantly.
- Lack of technology—employees want it but also afraid.
- Finishing department on low end of pay scale.
- Engineers, etc., use mill as training grounds for other jobs.
- Unskilled workers have little chance to improve situation.
- Lots of dead-end jobs.
- Lack of trust between management and work force.
- Workers treated like schoolchildren.
- Work load has increased.
- Problems are kept secret, not allowed to talk about them. When they are solved, management solves them without letting anyone know what is going on.

3. Unit trends (customer service):
   - Too much supervision.
   - Lack of say in how to do job.
   - No positive feedback.
   - Physically crowded.
   - Communication poor because of partitions in office.
   - Volume of paperwork so great that it must be spread among three people, information not kept together.
   - New paperwork dumped on people to fill up spare time.
   - Useless work being done.
   - Performance measures not fair, poor performance frequently results from

foul-ups in amendments, teletype messages, cross-referenced orders with different shipping dates not under the control of those being judged.
- No real system for getting work done, crisis orientation.
- No planning function.

4. Tasks
   - Special projects.
   - Dispatch trucks.
   - Process orders.
   - Schedule shipments.
   - Prepare bill of lading.
   - Maintain inventory levels.
   - Revise ship dates.
   - Prepare late shipment reports.
   - Correspondence preparation.
   - Delete orders.
   - Filing.
   - Quality alters.
   - Provide information to corporate inside sales.
   - Match tallies.
   - Responsible to ship pool trucks.
   - Sheet backlog report.
   - Direct and maintain order shipment status.
   - Compile truck yard check.
   - Transmit telecopies and teletypes.
   - Receive and distribute orders/amendments.
   - Tally edit.
   - Mark off report.
   - Prepare cutter orders.
   - Prepare rewinder orders.
5. Improvements

   Level 1 (departmental)
   - Shut door, stop people from using office as shortcut.
   - Put hold button on phones, install intercom/call commander.
   - Stop people from other departments from coming in and bothering customer service representatives when customer service representatives are working.
   - Regular breaks.
   - More filing cabinets.
   - Install word processor/provide training.
   - Do something about drafts in room.
   - Define work priorities.
   - Cross train.
   - Paint the office.
   - Get managers to provide all necessary information.

   Level 2 (divisional/PS&D)
   - Improve communication with sheet room.
   - Develop passwords for gaining access to required information.

(continued)

- More training on new process control system.

Level 3 (mill-wide)

- Interoffice porters.
- Get key punch machine.
- Improve overall training system.
- Stop unnecessary calls from different places within company when desired information is already in the system.
- Get five-day prior release for PM cycle (closed).
- Stop accessibility of all inside sales reps to system, establish coordinator.
- Direct amendments to area of concern.
- Make sure proper identification is on order changes.
- Appoint a trim coordinator.
- Give us 48-hour lead time concerning stock on hand.
- Make sure the inventory level information we are given access to is accurate.
- Improve communication on cycle changes.

Into the Trenches

By the next day it was obvious that Tom had made a mistake in not explaining the team building process. A rumor had spread that he and Doug were efficiency experts, that their ultimate objective was to eliminate jobs. At the beginning of the second meeting a worker stood up and announced that the Customer Service team wanted union representatives present from this point on. Tom apologized for his shortsightedness, then proceeded to explain the process in detail. He stressed the point that because the team effort helped increase productivity, a successful effort would most likely result in more rather than less jobs. The objective of the OP&D approach was to help improve the bottom line through better utilization of employee expertise, rather than through the reduction of labor costs.

Team members, however, were not convinced, and eventually Mr. Goins had to be called in. When the demand for union representation on the team was voiced, he said absolutely no. The unions had been briefed as to what was planned and had said they were willing to give it a chance. There was no need for a representative to attend.

The next step was to prioritize the level-one issues previously identified by the team, starting with those which would be easiest to address. The customer service staff, however, argued that some of the issues from other levels were also top priority ones (See Table 7.2). Tom said, "Fine. It's your team. Give it a shot. We don't, however, want to become immediately involved in a project

Table 7.2 Project Priority List for Customer Service

Priorities (not in order)

- Develop an adult atmosphere in office. (Level 1)
- Gain the necessary knowledge and expertise with the new process control system. (Level 2)
- Define job priorities and responsibilities and give access to the necessary information. (Level 1)
- Get two signs made to put on office doors (No Thru Traffic). (Level 1)
- Get additional file cabinets. (Level 1)
- Install a call commander or put hold buttons on phones. (Level 1)

that's going to take months. Our objective is to show quick results. We want to prove to ourselves that we can create change. We want to show management the value of the team approach."

Departmental communication was then selected as the number one priority project. Team members felt they frequently lacked information that they needed access to, and that their job priorities and responsibilities were often not well-enough defined. PS&D had two supervisors, the department head and his assistant. These two were isolated physically from the rest of the staff in their own offices and were frequently unavailable.

As an initial step, the team wanted to invite the department head and his assistant to the meeting. This was done immediately, but when they arrived no one spoke. Tom was forced to present the project list and to explain the first priority. When the senior manager asked what was wrong with department communications, someone finally said that the staff was tired of being treated like kids. The department head retorted that they were treated like kids because they acted like kids.

Tom guided the team into defining systemic ways to improve the situation. Allowing the personal issue to surface had been necessary, but allowing participants to dwell on it would not have been as productive as encouraging the team to address the problem as a systemic rather than a personality one.

Within an hour the group had generated a list of five ways to improve departmental communication. Also, team members had agreed to begin writing their own job descriptions. At this point the managers had to leave for another meeting. Following their departure, the hourly workers decided that the department head and his assistant should be invited to attend team meetings on a regular basis.

Priority Project 2

The second priority addressed was training on process control systems. The two systems involved were new. They had been installed by corporate experts. These people had then spent several days training Customer Service Department users. When these experts left, however, and the trainees actually begun using the new technology to schedule machine runs and integrate mill-wide operations, a great number of questions and doubts had arisen. Users admitted that at present they were frequently guessing. In other instances they used the new system, then repeated the procedure on the old one as a check. Finally, some users were simply ignoring the new system and depended entirely on the old one.

The facilitator selected for this team by Mr. Goins, a young engineer, had joined the second session. Her instructions were to sit and watch, to become familiar with the team's problem list and priorities and the way Tom facilitated the team's efforts. When team members began explaining their doubts about the process control system, however, she grew alarmed. The entire mill operation was affected by what customer service fed into this system. She asked, "Why haven't you called the corporate experts with your questions?" Some team members replied that they had been afraid to. Others said that they had tried to call, but, due to the experts' travel schedules, they had encountered difficulties in reaching people with the right answers.

The facilitator said she would take care of the problem immediately. Tom interrupted to tell her that problem solving was not her role. It was a team project. Team members should make the contact, but first, one of the ground rules had to be honored, the one about involving everyone directly affected. Who would that be? The team responded that everyone in the mill was affected. Tom asked where they thought they should start in terms of involvement. Team members said to start with their own supervisors.

The facilitator invited the two men to join the meeting once more. Both were greatly disturbed by what they heard. The department head demanded to know why staff members had not asked their questions when the experts were on site. Team members responded that they had not known the questions at that point. The department head said he would take care of the problem immediately. Again Tom said, no, it was a team problem now. The team would deal with it, but needed his support and input.

The department manager was extremely upset when he left. His assistant stayed and said that he would like to join the team if possible, that the department head did not communicate well with him either. That is why he had not been of more use to the hourly workers. If the team process could help change that one relationship, it would produce a great improvement.

Other key stakeholders in terms of this issue were the PS&D division superintendent and top-level management. The assistant department manager said he would discuss the problem with the division superintendent. Because team members were reticent at this point to address top-level management, not knowing what the reaction would be, the new facilitator volunteered and was given that responsibility.

The next step was to generate project action steps. Tom asked the team to think in terms of an ideal scenario. What would members want to happen ideally? The answer was that an expert would be flown in again, not to give a presentation this time, but to be available for as long as was necessary to answer questions and to help users develop the necessary familiarity with the system.

Unless other stakeholders disagreed, then, the first requirement was for someone to call corporate and request a return visit by an expert. Team members balked. They said they did not have the authority to make such a call. Tom insisted. It was their project. If it was going to be done, someone in the room would have to do it. Finally, one of the older employees volunteered. The facilitator said she would help the volunteer reach the right person. The employee asked what she should do if no one would take her seriously. Tom answered that if the other mill stakeholders agreed to what was being done, the ground rules assured their support, right up to the mill manager.

A second action step defined by the team was for customer service personnel to begin compiling a list of questions to be sent in advance to the expert picked to visit so that he or she could prepare.

By the end of the session the team members were at least curious and seemed willing to give the process a chance.

The last stop for the OP&D consultants was the mill manager's office for an exit interview. They said nothing about the project the PS&D team was working on, but instead suggested that a long-range planning exercise be started immediately to provide a frame of reference for problem-solving team efforts. Mr. Goins, however, said it was too soon. His direct reports were not familiar with the idealization approach that was to be used. They needed to talk about it. They also needed to develop an understanding of the team problem-solving process. Finally, they needed to start solving some problems themselves, instead of him always taking the lead. This last comment was important. It showed Mr. Goins' realization that if the process were to succeed he had to set the example by backing off and giving his direct reports more authority.

Topics for Discussion

1. Why did the new Vice President of Quality pick the model he picked?
2. Why did OP&D feel qualified to question this model?
3. What resistance did OP&D run into while looking for a site to test its model?
4. What mistakes did the OP&D consultants make from a systems perspective when starting the PS&D team?
5. What were the keys to success in keeping this first team going?
6. In what ways has top-level management's commitment to the process been tested at your organization?

8 Maintaining the Momentum

After Reading This Chapter You Will Know

- How teams learn the value of the ground rules.
- How facilitators were trained on-the-job.
- Why teams decided to release their meeting minutes.
- How the productivity of the team network snowballed rapidly.
- What caused the mill manager to buy into the process.
- How the interactive planning process was started.

The Mill Manager Comes Through

Doug Ferguson and Tom Roberts returned to New York City. The plan was for OP&D consultants to spend 2 to 3 days each week at the mill during the initial phase of the process and 2 to 3 days at corporate headquarters. Almost immediately, however, they began receiving calls from customer service team members. Things were not going smoothly. The department supervisor felt he had been put on the spot without proper warning. The PS&D division superintendent had supported him and had carried his complaint to the mill manager.

The department supervisor had also told the team that if it wanted to talk with him, two representatives should make an appointment and come to his office. He would then take whatever they said under consideration and get back to them if it made sense. Two team members, according to phone reports, had become physically ill following Tom and Doug's departure. They were afraid of reprisals. They felt as though they had been put into a dangerous position, then deserted.

The next visit to the mill occurred during the week of February 12, 1986. After talking with Mr. Goins, Doug and Tom went to see the department head of Customer Service and his assistant, first to listen, and then to review their role in relation to the team. Earlier in the week Mr. Goins had made clear to them and to their superintendent that the team building process was not going to go away, that they would have to learn to live and work with it.

During this session Doug and Tom suggested that when team issues were discussed, the two supervisors would meet with the entire team rather than just representatives. They suggested that such a meeting be held in a more neutral location. Finally, they suggested that the supervisors not let the team dump problems into their laps. That was the old way. The team now had to be forced to take responsibility for discovering the best solution and for implementing the action steps outlined.

Tom and Doug then met with the team and reviewed its conversations with the department supervisors. One member wondered out loud if it was worth the pain. The supervisors were then brought in and the stakeholder ground rules were covered again.

The next item on the meeting agenda was to discuss the progress of ongoing projects. Owing to the confrontation, little had been accomplished. With the managers still present, action steps and assignments were reviewed. Because nothing more could be done on the first two priority projects until the current action steps had been completed, the team decided to move on to its third priority, the redesign of the flow of work through the office. It began identifying and seeking reasons for bottlenecks in this flow. One reason offered was job security. Employees were still traumatized by earlier efficiency increasing studies. They felt that they could improve their chances of surviving the next cut by hoarding tasks and information.

Tom explained that two approaches to improving the bottom line existed. The first was decreasing costs. In such cases getting rid of employees was too frequently the easiest alternative. The second was finding ways to increase work force productivity. The team was the nucleus of this second approach. If the team building effort produced the expected results, employment security should increase rather than decrease. In order for the team approach to work, however, members had to help each other, to share information and skills.

Team members then surprised Tom by suggesting to the department head that his assistant be allowed to attend the daily meeting between himself and the PS&D superintendent. This way he could help keep the team better informed. Both men seemed startled. Team members pressed their attack, pointing out ways the change would make his job easier and them more effective. Eventually, the department head said that he would see. It was also decided that the assistant should become a permanent team member. He could help keep the department head and division superintendent informed of team activities.

One final issue addressed at this session was team minutes. The ground rules stated that the team could decide to keep them private or to distribute them. If it picked the latter course, it had the right to decide who should get them. Tom suggested that Mr. Goins be a major consideration. He had initiated the process and had protected the team. It was important to show him that the process was producing results. After much discussion and some pressure from Tom, the team decided to distribute its notes to Mr. Goins, but to no one else at this point.

Team Two

During that same week one of the roll finishing teams was formed. Workers in this function rewound paper from the large rolls coming off the papermaking machines, cut it to size, and made up the smaller rolls that met customer specifications. Then they wrapped these smaller rolls and prepared them for shipment.

Owing to the size of the roll finishing operation and the number of people involved (180), Tom and Doug decided to form two teams —C/D shifts and A/B shifts. This decision contradicted the belief that all shifts in a function should be represented on one team. Doug and Tom, however, wanted to involve as many people in the effort as possible. The decision gave them approximate 24 team members instead of 12. Also, the team building process was flexible. It was a learning experience for everyone. Nothing was carved in stone. If the arrangement did not work, they could revert to a functional breakdown.

When Tom and Doug met with shift foremen from roll finishing to introduce the process, the first comment they heard was, "This team concept is full of. . . !" After that, however, things went well. The foremen agreed that they had to deal with too many small, time-consuming, day-to-day problems, that if they could get the hourly workers involved in handling some of these it would give them time to address larger issues. Everyone decided to give the team approach a chance. If it proved itself, fine. If not, it could be stopped by a wave of Mr. Goins' hand.

The idea of having two teams instead of one was presented. Meetings could be held at shift change-over, representatives from the out-going shift staying an extra hour. The foremen asked if those staying the extra hour would be paid overtime. Doug repeated Mr. Goins' decision that they would. The foremen said that the alternative of having all four shifts represented on one team would never work. Those sleeping or not working that day or night would never come in.

In the first start-up session with roll finishing shifts C/D Tom explained:

1. What the team building process was for.
2. How it worked.
3. The hoped for results.
4. How it probably would improve employment security.
5. The ground rules.

A search conference was then run. The community trends identified were generally the same identified by customer service. It was decided to skip identification of corporate trends completely to save time and to avoid the confusion of separating them from mill trends. The mill trends identified were also similar to those identified by the customer service team. Most of the first meeting was spent identifying unit trends and responsibilities.

Doug ran part of the second start-up with Tom supporting him. He guided the team through its definition of desired improvements, the breakdown of these improvements according to the team level that would have to address each, and the definition of priorities (see Table 8.1). He also introduced a simple technique for defining project priorities. Each team member would call out his or her top three. Doug would give the choices three, two, or one point. The priority that ended up with the greatest number of points would become number one, and so on. If a tie occurred, there would be discussion, then another vote.

Before heading back to New York, Tom and Doug visited the mill manager again. He had by this time received session notes from customer service and was surprised that staff wanted more training on the process control system. He said he would make the necessary calls immediately. Tom and Doug had to work hard to convince him not to jump the process. Mr. Goins eventually shook his head and said that it seemed like the long way around; but, on the other hand, if the team had not brought the problem out it could have festered for months. This was, indeed, a new way of doing things, but he would try.

Table 8.1 C/D Shift Roll Finishing Team's Improvements/Priorities

Level 1:

- None.

Level 2:

- Reorganize training.
- Define position responsibility clearly.
- Balance work load.
- Standardize procedures between shifts.
- Stop competition between shifts.
- Proper equipment and proper quantities.
- Each area properly equipped.
- Shifts share equipment.
- Improve communication with order department.
- Designate inventory area in warehouse.
- Need short boom tow motor.
- Need more dependable equipment.
- Relocate tally shack.
- Need tally printers in all shipping shacks.
- Need clean up people to increase productivity.
- Need recognition when job well done.
- Create an adult, trusting environment.
- Reinstitute position of head roll finisher.
- Plan ahead to match available transportation with production.
- Get skill saw to cut boards to brace trucks and rail cars.

Level 3:

- Reestablish chain of command.
- More recent reference sheets.
- Coordinate production of roll sizes.
- Replace large rewinder.
- New roll wrapper.
- Assign maintenance to department from 4 p.m. to 8 p.m.
- Need noise control for shredder.
- Reduce overruns when making rolls.
- Make sure we get accurate input cards.
- Improve communication and cooperation between winder and roll finishing.
- Purchase and install banding machine.
- Develop unit strictly for the maintenance of PS&D tow motors.
- Make crew concept work or replace it.
- Control inventory levels, think just-in-time not just-in-case.
- Curtail last-minute changes in shipments.
- Supply trucks just-in-time instead of backing them up in the yard.

Priorities

1. Reorganize training.
2. Create an adult, trusting atmosphere.

(continued)

3. Balance the work load.
4. Designate inventory areas in the warehouse.
5. Equip each area properly.
6. Reinstate head roll finisher.

Tom explained that when only one or two teams were involved and one or two projects, the process was, indeed, inefficient in terms of time. But when 50 or 60 teams were eventually working on several hundred projects the payback in terms of time and efficient use of resources was great. A little patience at the start would produce tremendous benefits in the long run.

The second issue discussed at this meeting was employee security. So long as employees suspected they might be fired tomorrow during an efficiency purge, generating the necessary degree of commitment to the process would be difficult. Though corporate was constantly pressuring the mills to reduce numbers, was there anything Mr. Goins could do to alleviate his people's fears? They respected him as a man of his word. If he could come of with something. . . . Mr. Goins said, yes, he understood the problem and would work on it.

Finally, Mr. Goins said that there were projects that needed to be started in PS&D. Did management have to wait for the teams to identify and choose to work on them? The answer was no. Management was free to start any project it wanted, to build any task force, include any QIP team members that it wished. The improvement related problem-solving team network approach being developed by OP&D consultants was *not* supposed to replace normal operational problem-solving procedures, but to complement them.

At the same time, however, management could not take over the quality teams, could not force them, at least initially, to do jobs it thought important. The teams *had* to be allowed to identify their own priorities if the desired degree of ownership and commitment were to be generated. One of the things that made this arrangement easier to accept was that more often than not team priorities matched management priorities.

The Third Week

By the following week, a lot was going on at the mill. Mr. Goins had found space for team meetings in an old training building that sat away from the mill. The OP&D consultants had requested such a space, which was free from distractions, which the team could eventually claim as their own. It was a bit of a walk, but this provided a small break—a transition period for workers coming from their jobs.

The facilitator ran the fourth meeting of the customer service team with an OP&D consultant sitting in support. The meeting was spent reviewing action steps in process on priority projects and identifying new projects (see Table 8.2). Team members were beginning to believe that they could make things happen. For example, the assistant department supervisor had been invited to attend morning meetings with the department supervisor and the division superintendent.

Much of the fear was gone, replaced by cautious enthusiasm. It was quickly becoming obvious that most of the actual work on these projects would be done during the week. Team meetings would be mainly for review of ongoing activities and planning of future ones.

Priorities picked by the C/D shift roll finishing team included:

1. The training and cross-training of shift workers.
2. Reinstitution of a head roll finisher, a position cut out as a result of the corporate efficiency thrust.
3. The availability of tools to all sections and shifts.

In terms of entry number 2, Tom pointed out that according to the ground rules the team had to show the reinstitution of this position to be economically beneficial to the mill. It had to develop a believable cost-benefit analysis balancing the salary of the extra position against the loss of efficiency caused

Table 8.2 Customer Service Session Four Notes

1. Make process control system training mandatory for everyone in the department.
   - Bring in trainer from corporate.
   - Make upper-level management aware of need.
   - Work with each other to improve understanding.
2. Put order in for No Thruway signs for doors.
3. Develop justification for floater to cover illnesses, vacations.
4. Discuss paperwork excesses with stakeholders in other departments.
5. Develop individual lists of training we need, training we can offer each other.
6. Find a way to get a copy of roll shipping information earlier.
   - Should this be retrievable from system.
   - Window needed at corporate.

by others on the shift having to handle that position's responsibilities in addition to their own. The team said that such an analysis would be difficult to put together. Tom said, true, but this was how the game was played. The ground rules also said that any project proposed had to have the approval of all affected, and upper-level management, as such a stakeholder, was not going to reverse a corporate staff decision without good reason.

In terms of entry number 3, Tom pointed out that the availability of tools for crew members was always a problem in primary manufacturing operations, that when made available they tended to disappear. The only way team members were going to convince management to provide more was to convince it that the team or the shifts represented could and would take responsibility for the tool not walking.

The team decided to focus on this issue during its next meeting. It also decided that it would be to the team's benefit to release meeting notes to anyone interested, from the mill manager on down. The major mill-wide problem, in the team's opinion, was lack of appropriate communication, and this was one way of addressing it.

Doug ran the entire start-up exercise for the third team formed—roll finishing shifts A/B—using the same introduction and search conference format used with C/D shifts. This team, possibly because members had talked to their mates on C/D, became involved without any hesitation. By the start of the second session, it had already defined its priorities and was set to start solving problems. While many of the level two and three issues identified were the same identified by C/D shifts, the priorities, with one exception, were at least worded differently.

Instead of beginning the problem-solving process with an analysis of its first three priorities, Tom suggested, as an experiment, that the A/B shifts team idealize. It should start by identifying the characteristics each system ought to possess ideally, rather than its faults. The results of this approach can be found in Table 8.3.

The team's fourth priority, knowledge of the new process control system was the result, mainly, of fears about the new system eliminating jobs. Instead of idealizing, members generated a list of questions for the mill manager and the expert from corporate to answer while the expert was working with the customer service staff (see Table 8.4).

The issue of releasing meeting notes was brought up at that point. The team, following the lead of C/D shifts, decided to release them to anyone interested, as long as individual comments were not included and opinions and decisions were presented as those of the team as a whole. Tom explained

Table 8.3 A/B Shifts Roll Finishing Team Definition of Ideal Characteristics for the Systems Addressed by Its First Three Priorities

1. Pressure vs. problem -solving
   - Should have one boss, reestablish chain of command.
   - First ask operators what the problem is, then solve it together.
   - Do not blame us, or just tell us, work with us.
   - Adult atmosphere.
   - Management should stop assuming we are goofing off, that we do not care about our jobs.
2. Training:
   - Should be more on the job.
   - Should be led by experienced operators.
   - Properly trained back-up people should be in place.
   - Consistent training schedule should be developed.
   - Proper training should be received before starting job.
   - Should be on going.
   - Training program should be flexible, some learn faster, some need more training than others.
   - Same trainer should be used over time.
   - Instead of using spare time to clean up, paint, etc., use it to train.
   - Stop cancelling vacations because there are not enough trained people.
3. Crew concept:
   - Rotation schedule should be consistent—rotate every four weeks.
   - Rotate one group at a time—not the whole shift simultaneously.
   - There should be more consistent movement upward.
   - Train people before promoting them.
   - Redistribute key people with the necessary training so that more of us have access to them.
   - Let workers know how the decision as to who rotates is made.

Table 8.4 Questions Concerning New Process Control System

- How many jobs will be affected?
- Which jobs will be affected and how?
- What is going to happen to the people whose jobs are affected?
- How does the system work?

to the team that this need to protect individual team members and their opinions was one reason decisions were reached through consensus rather than vote. A second reason was that until everyone accepted it, the most workable solution had not yet been achieved. A third was that it was more efficient to spend the time up-front generating the required level of commitment and understanding than it was to try to do so during the implementation phase. Again, in terms of reasons two and three, the ground rules made it impossible for teams to act otherwise.

From that point on, all new teams decided to release their meeting notes to anyone interested. A distribution list was developed including the mill manager, his direct reports, and everyone in the management hierarchy of the individual team. Copies were also made available to the hourly workers represented by the team.

Tom, Doug, and Joyce next met with foremen from sheet finishing to brief them and to ask them to select team members for the following week's start-up session. The sheet finishing staff was small enough that one team seemed appropriate. The foremen already knew much of what was going to be said from the grapevine, and their meeting with the OP&D consultants went smoothly.

During the squad's weekly exit meeting, Mr. Goins announced that the process should go mill-wide. He had received all team notes thus far generated and liked what he was reading. He was impressed that positive changes were already occurring. He had thought it would take much longer to get the hourly workers involved. Mr. Goins also said that a session should be set during the next visit to begin the long-range-planning exercise with his direct reports. Finally, the union presidents had called him and asked for a presentation. This should be scheduled.

Week Four

The fourth week of team building, that of February 26, went according to plan. The customer service team had received approval for its project action steps from all stakeholders including the mill manager. They had, between meetings:

1. Contacted corporate about the need for more training on the process control system, sending along a list of questions. A trainer was scheduled to come in the next week and to stay as long as they wished.
2. Sent a representative to talk to the mill manager about the need for a floater, one of the positions lost during the last efficiency study, presenting a rough cost-benefit analysis.
3. Talked to a corporate expert about further automating their office, the types of equipment that would help improve efficiency, and how it

could be integrated. The expert was helping them over the phone, and said he would visit the mill soon.

Roll finishing C/D cancelled their week's meeting. Roll finishing A/B had decided to make the reorganization of the warehouse its first priority project. During A/B's meeting a problem analysis was done with a few suggested improvements thrown in (see Table 8.5). The team decided that the most important action step was to encourage upper-level management to give one individual responsibility for running the warehouse and for setting and enforcing policy. Currently, a number of supervisors had developed their own individual policies, especially concerning the allocation of space, and these policies often conflicted. The team decided to hand-distribute its meeting notes immediately to all stakeholders as a means of making sure they became aware of the seriousness of the problem and of the team's suggestions.

A/B shifts team members also decided they needed to meet with representatives of the C/D shifts team. A great number of the issues which interested them involved all four shifts. When asked, however, if they wanted to combine the two teams, the answer was, no, they did not want to dilute their effort. They wanted to continue working on the warehouse and other priorities by themselves. A task force including representatives from both shifts should be formed to deal with cross-shift issues.

Table 8.5 A/B Shifts Roll Finishing Team Session Three Notes

Warehouse Problem Definition:

1. You cannot stack orders together because there is not enough room.
2. Area too strung out, loading docks are at north end and split.
3. Cannot find orders because they have been stacked wherever room can be found.
4. Orders sometimes hidden behind other orders.
5. Cannot find orders because computer lies, says we have things that we do not have, does not tell when something has been shipped.
6. Too many orders have to be stored, not enough shipped immediately.
7. Poor communication with paper machines, customer service.
8. Computer goes down almost daily.
9. Too much bad paper coming down, someone up the line is trying to meet his production quota no matter what.
10. No one is in charge of the warehouse, everyone you go to has a different answer.
11. Need 24 hour switching to move trucks in and out of loading docks.
12. Need another tow motor driver to haul from one area to another.

Sheet Finishing

Doug and Joyce ran the two day start-up for the sheet finishing team, which included representatives from all shifts because the function involved only 50 hourly employees. As decided, participants were assigned by supervisors. Those off shift who had been forced to come in were unhappy and wanted to know if this added duty was permanent. It was explained that the first five meetings were obligatory. If, at that point, members decided they did not want to continue, replacements would be found. In general, however, the process was greeted with enthusiasm, and the team identified 38 projects it wanted to work on.

The presentation to the heads of the union locals went well. Mr. Goins attended and Tom unexpectedly volunteered him to explain the purpose behind the team building process, how and why it worked. Mr. Goins did a good job, ending with, "I'm learning, too, a lot more than I thought I'd be." It was obvious that even after this short period of time he was beginning to develop a gut-level feel for what was involved and the value of it. The union heads seemed pleased and made comments like, "This is long overdue," and "This is good."

Finally, Tom introduced the long-term planning phase of the process to Mr. Goin's direct reports and several other managers who had been added to the group to make up a lead team. The plan was for Mr. Goins to facilitate this team. He started by giving the same presentation he had given to the union heads. There was, however, no response. Tom introduced the concept of idealization. Again, there was no response. Only one person other than Mr. Goins and Tom spoke during the entire session.

Back in Mr. Goins' office Tom suggested that either Doug or he assume the role of facilitator until momentum had been gained. He also suggested that Mr. Goins miss the next several meetings for the same reason that foremen and supervisors were asked to stay away from the initial meetings of problem-solving teams. Mr. Goins accepted the recommendations, though he was obviously upset.

When the PS&D squad returned to New York, it received an unexpected surprise. One of Ron's responsibilities, as department head, had been to make sure people at corporate understood and accepted what was happening at Louisiana. Almost immediately, however, he had run into trouble. The Vice President of Human Resources had said such efforts were the Quality Department's responsibility, not OP&D's, and that he was terminating the OP&D project at the Louisiana Mill. Ron had called Mr. Goins with this information. Mr. Goins, in turn, had immediately called someone or some people on the operations side of the business. They, in turn, had convinced the Vice President of Human Resources to leave the project alone.

An OP&D staff meeting was held and it was decided that the best defense against such misunderstandings was to get as many key people from corporate down to the mill as possible to see what was going on. A list of critical corporate stakeholders was drawn up. Ron immediately began making appointments to talk with these people and to invite them to visit the mill and form their own impressions.

Week Five

During the week of March 5, the hourly PS&D maintenance team and the joint roll finishing team were formed. The maintenance start-up was run entirely by Joyce with Doug sitting in support. Some of the old-time maintenance workers were extremely skeptical. One eventually said, "This thing will last 6 months, then it will be gone, like everything else that corporate sends down. Somebody up there is just trying to get a little attention." Two other older team members spent the entire first meeting staring blankly at the wall. The younger workers, however, got involved quickly. One thing everyone seemed pleased about was the fact that maintenance personnel had been given a team of their own instead of being made part of other teams.

Tom started the joint A/B shift and C/D shift roll finishing task force. No one knew what to expect. The task force began by listing problems that would affect all four shifts, then prioritizing them. Its top three priorities were the warehouse, training, and tools. This confused things even more. In terms of the warehouse project, for example, who was going to take the lead, the task force or the A/B shift team?

Tom encouraged the joint team to reconsider its priorities and to focus, instead, on finding ways to improve cooperation between shifts. The task force agreed and started by listing things the individual shifts could do to improve relations. The list included:

1. Clean up after themselves.
2. Share tools.
3. Stop competing.
4. Stop leaving work undone.

Customer service and the C/D shift roll finishing team had, by this time, completed projects. Customer service had:

1. Acquired additional filing cabinets.
2. Made up and posted the desired No Thru Traffic signs on its office doors.
3. Begun its process control system training with the expert from corporate.
4. Received an on-site presentation from a corporate office automation equipment expert concerning what existed and what would be most appropriate for their situation.
5. Received missing manuals for the process control system.

The C/D shifts team had:

1. Finished an inventory of tools the shifts had access to and needed, then developed a system for safeguarding them.
2. Developed a system for returning damaged rolls to the rewinders.

The PS&D department now had three facilitators. All were young engineers. The facilitators' responsibilities were drafted by the OP&D consultants and reviewed by the new facilitators (see Table 8.6). It was also decided at this point that teams should choose their meeting times. Facilitators would be responsible for arranging their schedules so that they could be available. The teams, however, decided that their facilitator's work-load as well as their own should be taken into account when making this decision.

Table 8.6 Facilitator Responsibilities

1. Make sure ground rules are understood and obeyed.
2. Have team review prioritized projects and action step taken.
3. Have team discuss new action steps to be taken.
   - Make sure responsibilities are assigned to team members.
   - Make sure timetable established.
   - Have team discuss desired results.
4. Help teams define and gain access to desired resources.
5. Have team discuss new projects suggested by employees represented and add to list.
6. Have team define new priorities when projects are completed or dropped.
7. Make sure someone takes notes.
8. Schedule next meeting and write on calendar.
9. Set up system for rotating membership.

The Mill Manager Buys In

This time the end-of-the-visit meeting with Mr. Goins was an especially lively one. He said that he had to continually fight the urge to get directly involved, to solve the problems himself. An example was the inventory problem and the desire expressed by the roll finishing C/D shift team to switch to a just-in-time system instead of constantly stockpiling. Mr. Goins said that the answer to this one was simple. Corporate policy made the Louisiana Mill responsible for maintaining specific inventory levels, not for meeting orders.

The problem, as he saw it, was not that the policy was bad, but that the team did not understand it. In fact, after looking through team notes Mr. Goins said he had realized that at least 70 percent of the issues raised by the teams were partially the result of a lack of appropriate understanding. It was obvious that communication and education had to improve mill-wide. He had not realized that the hourly workers were truly interested in knowing these things.

In his own case, he could not believe he had just found out that no floor plan or agreed upon allocation of storage space existed in the warehouse. He was also surprised to learn that team members thought no one was in charge. A supervisor, indeed, had the warehouse as one of his responsibilities.

When Mr. Goins was informed that the involved supervisor had been contacted and had reported that his decisions concerning the warehouse were constantly being overridden by superiors, Mr. Goins asked if he was allowed to get involved as a stakeholder. The answer was yes. The C/D roll finishing team had defined the problem as members saw it, then distributed it with ideas to those who had to play a role in the improvement effort, including himself. There was nothing more the team could accomplish without his intervention. If Mr. Goins would inform the originating team of his intentions and involve it as a stakeholder in the improvement effort, everyone would probably be pleased.

One final observation was made before departure concerning Mr. Goins' comment about the need for improved communication. The team network being formed was, in essence, a communications vehicle. The ground rules ensured this. It forced focused and well-integrated communication in all directions concerning the issues addressed. The main purpose of the team building process was not actually to generate improvements in the operation, but to encourage a new openness in the mill management culture. Once people started communicating sufficiently, the improvements would take care of themselves.

Week Six

Before the PS&D squad's return on the week of March 12, Mr. Goins had seen to it that one supervisor gained full responsibility for and the authority to run the warehouse. He had also formed a task force to allocate the space. Representatives of all the PS&D Department quality teams were on the task force.

The agenda for the week of the 12th included:

1. A meeting with paper machine foremen.
2. Starting the lead team exercise.

The visit was relatively uneventful. The six PS&D Department teams met on schedule. OP&D consultants sat in, but contributed little. The facilitators by now understood their responsibilities and were gaining confidence in their new roles. The PS&D maintenance team had finished its list of priorities and broken them down by level. It had also begun work on its first priority problem—misuse of the work order system—by generating a list of the system's desired characteristics and another list of questions that needed to be answered concerning it (see Table 8.7).

The lead team met without Mr. Goins for its second session. The purpose of this session was to begin the idealized design exercise. The OP&D consultants had decided that this exercise should also back into the mill so as not to preempt hourly team activities. Shipping would be the first function addressed. Doug, as facilitator, would follow the standard idealized design steps:

1. Define the idealized mission of the shipping function in terms of all key stakeholder groups.
2. Identify key systems in the shipping area.
3. Define ideal characteristics each system *should* possess.
4. Identify systems design elements that will allow the function to gain these ideal characteristics.
5. Identify stakeholders the design must be discussed with and approved by before implementation begins.

While Doug explained the process and the logic behind it, no one spoke. Finally, Doug asked what the problem was. "Shouldn't we wait for Mr. Goins to arrive before starting?" was the response. When told he was not coming, the

Table 8.7 PS&D Maintenance Team Work Order System Project Desired Characteristics and Questions List

Characteristics:

1. Everyone should be allowed to write work orders.
2. Foreman should sign off on all work.
3. Everyone should be trained to write work orders.
   - Foreman should be trained by maintenance planning.
   - Foreman should train shifts.
4. Eventually, work should not be done without work orders.
5. Two key pieces of information are necessary.
   - What is the problem specifically?
   - Where is the problem located?

Questions:

1. What should be written on work orders?
2. What is the procedure for submitting?
3. How should training on filling out and submitting work orders be organized?
4. When should work orders be written?
5. What should the approval level be?
6. How should an incentive system be set up to get work done as efficiently as possible?
   - The hourly wage system, in effect, encourages workers to draw the work out.

group seemed somewhat surprised. Doug pressed the issue, asking why he had to be there. That question brought rueful chuckles and a comment to the effect that Mr. Goins made all of the decisions.

Doug continued pressing so that the rest of the meeting was spent discussing the manner in which the mill had been run up until now and the way it should be run. Doug listened, asked questions, and ended with the observation that he had never dealt with a CEO or facility manager who learned more quickly or showed greater support for the process in its early stages than Mr. Goins had. If Mr. Goins was not seriously interested in change, why was he doing these things?

No one had an answer. Doug continued by saying that the idealization exercise was a safe way to test Mr. Goins' commitment. As a team, rather than individuals, they could identify ways to improve the shipping system. If Mr. Goins liked the suggestions, fine. If not, he could end the exercise without anyone getting hurt. This was one of the values of the team approach. While the wider range of inputs usually made the product richer, no one's name or everyones name was on it so that no one could be scapegoated.

The superintendent of PS&D expressed concern about all these people from other parts of the mill coming in to redesign his operation. Doug asked if he believed they really understood what went on in his area. The answer was no. Doug asked if he believed the development of such understanding would benefit his operation. The answer was yes. Doug then suggested that he view the exercise as an opportunity to educate his peers. At the same time, because decisions were made by consensus, no changes would be allowed without his approval.

The lead team met again the next day. Much of the tension had disappeared. The idealization exercise began immediately with identification of key stakeholders to the shipping function and the development of a mission statement saying what the function should do or provide for each. Key systems were then identified. One, the truck dispatching system, was singled out and the characteristics it should possess ideally defined.

Week Seven

The OP&D consultants' agenda for the week of March 19 read as follows:

1. Third session for lead team.
   a. Continue PS&D idealization exercise, focus on the shipping function.
      - Develop scheduling system that allows turnaround time of 24 to 48 hours.
      - Idealize the system for choosing carriers.
      - Idealize the system for notifying carriers.

   b. Identify stakeholders who must change the way they do things, who must contribute to, or be consulted on, the lead team design.
2. Give start-up presentations to the paper machine department's management corps, decide who will sit on the hourly teams in this department, schedule the first session for these teams, identify facilitators.
3. Sit in on at least some PS&D problem solving team sessions to make sure teams focus on and complete priority projects before starting others.
4. Try to get calendars put up in the meeting rooms for facilitators to schedule team meetings on.
5. Worry about too many things going on simultaneously in terms of meeting room space and OP&D resources.

Doug met again with the lead team. The mission of the shipping function had previously been defined as on-time delivery. The team continued working during this and the following week on the characteristics shipping should have ideally to fulfill its mission. The results of their efforts are found in Table 8.8.

During this part of the exercise two things happened. First, the managers of other functions began asking questions and making statements like, "I didn't know you did it that way." The second was that the team realized it had to include other stakeholders, specifically trucking company schedulers and drivers, supervisors and hourly workers from shipping, and the customer service team to understand whether the desired characteristics could be designed into the system or not. The team was encouraged to finish a strawman based on its own ideas, then to present it to the other key stakeholders so that their suggestions could be incorporated.

The hourly teams visited were all immersed in projects. Customer service was pricing word processors as well as the equipment necessary for their recommended phone system modifications. Roll finishing C/D had developed a secured storage space for tools and spare parts and was waiting for the tools they had ordered to arrive. This team was also developing a form on which the reasons for roll rejects would be recorded. A/B was taking the lead in the warehouse reorganization. The joint roll finishing task force was working on getting customer feedback on the condition of the rolls upon delivery.

In general, things seemed to be going well.

Table 8.8 Characteristics of Idealized Shipping Function

1. One person coordinating/supervising the effort per shift.
2. Up-to-date records of truck supply and paper inventory.
3. Low-cost carrier used.
4. Trucks loaded from dock closest to paper location.
5. Loaders have no trouble locating truck to be loaded.
6. Bracing of trucks/rail cars done efficiently.
7. Minimal reloading.
8. Shipping department maintains close contact with dispatcher.
9. Centralized shipping department.
10. One stop for carriers.
11. Computerized control of product location.
12. Computerized control of carrier pool and docking facilities.
13. All loaded trucks/vehicles inspected/verified to meet the mission of the function.

Topics for Discussion

1. Why was it possible to start teams without up-front training?
2. What were the fears of the first PS&D team and how were they dealt with?
3. How did the mill manager deal with his changing role?
4. How do you think top-level management in your organization would react to full employee empowerment?
5. What were the fears of lead team idealized design participants and how were they dealt with?

9 Success Then Setback

After Reading This Chapter You Should Know

- How facilitators dealt with team defensiveness concerning their projects.
- How the mill dealt with a manager who could not change.
- Why the mill manager was the key player.
- The step-by-step progression of sample projects.
- How the mill kept all employees informed as to process progress.
- How the mill dealt with attempts by the corporate quality improvement department to change the model.
- What happened when corporate abruptly disbanded OP&D.

Payoff Time

Tom and Doug were getting tired. Since the project's start, they had spent three days of each week at the mill. Two things were decided. First, now that Joyce was trained Tom, Doug, and Joyce could rotate, only two of them flying down each week. Second, someone at the mill should be incorporated into the OP&D consulting group and trained so that hourly team members could have constant access to an on-site representative capable of addressing their questions and problems.

The head of the mill Human Resources had too many other responsibilities. The squad's next choice was Sid Franklin, who was in charge of mill-wide training. Sid had done a good job of cleaning up and reorganizing the building Mr. Goins had donated. He had begun attending the meetings of a variety of hourly teams and had volunteered to become a facilitator. Perhaps most important, however, Sid had worked at the mill for 23 years, knew everybody, and was liked by everybody. Sid seemed pleased when asked, and Mr. Goins immediately okayed the suggestion. Sid's first assignment was to design display sheets to show each team's projects and their completion dates. The sheets would be hung in the entryway of the new meeting building.

During the week of March 26, Tom, Doug, and Joyce each brought a paper machine area team up. Sid attended most of the start-up sessions. The following two weeks were spent nurturing the new teams, working with the new facilitators, and striving to better integrate the model. During this period several critical issues were addressed. The first had to do with the lead team design exercise that encompassed the customer service function. The customer service team was upset. Once again bosses were planning their function's future without asking for input. This went directly against the process ground rules.

It was explained to customer service that the lead team was developing a strawman which they would then be asked to critique and to help modify. Customer service was not pacified. The issue of employee security again reared its ugly head. Was this a plot to get rid of jobs? By the time customer service was allowed to contribute to the design it would be too late. The job security issue had come up in other teams as well, the question being, "Will all these improvements make the operation so efficient that we lose our jobs?" It was becoming increasingly obvious that the OP&D consultant's assurances would not suffice. Something more concrete had to be offered by someone with real power.

The lead team decided to meet immediately with the customer service team and share its ideas. The first step in this direction was to generate a document that turned the already existing PS&D mission statement and characteristics into questions. For example, the mission statement of on-time deliveries was turned into the question, "Should the mission of the shipping function be simply to ensure on-time deliveries?" In terms of characteristics, examples of the new format included, "Should the scheduling clerks cross-train so they can support each other?" and "How can PS&D ensure 24-hour notice to carriers?" Of course, many of the issues being addressed by the lead team were also on the customer service team's list, and many of the conclusions the two teams had reached independently were similar.

Almost as soon as the two groups began talking, the fears of most of the customer service representatives dissolved, replaced by a feeling of pride that they were helping instruct the lead team, and were being dealt with as equals while working on a joint project.

A second key issue which took shape during this period was the negative impact the superintendent of PS&D was having on the process. He was an old-style manager who had been with the company 30 years. He was reported to run his division from his shirt pocket which contained a stack of cards with notes on them. He was a diligent worker who spent a great deal of extra time at the mill because he felt the need to control every facet of his operation. He continually usurped the responsibilities of his supervisors and foremen. He rarely passed any information from above on down to them. His attitude and actions had for years adversely affected the moral of the work force in PS&D. He had been discussed as a major impediment to change by every team formed in that area. The general feeling was that until he had been dealt with, little else could be accomplished.

Tom and Mr. Goins talked to this man, first individually, then together, trying to convince him that a change of style would benefit him as well as his staff. The job of running PS&D was too big for any one person. He would kill himself trying to do everything. During each of these sessions, the superintendent listened attentively and seemed to agree. Following each session, however, he reverted immediately to his old habits. Finally, Mr. Goins said, "You've got to change, George. If I can, you can. I know it's hard, but you got to make a decision." The superintendent shook his head, "I don't think so, Mr. Goins. It's been too long. I'm used to doing things my way. Change just doesn't make much sense to me."

The superintendent eventually took a job with another company. He was the only casualty of the process. Other, mainly older managers had trouble accepting the new culture. Most of them, however, sat back and waited quietly for it to go away. They were not confronted by the OP&D consultants or by team members. They were called on to provide information to teams and to function as stakeholders to projects affecting their part of the operation if they wanted to have a say, but nothing was forced. As process momentum built and improvements piled up, most of these men eventually changed their minds.

The belief behind this approach was that the process had to prove itself. If it produced results that improved their quality of working life, the doubters, as reasonable people, would come to understand and accept its value. The only requirement was that while they were waiting to be convinced, they, like everybody else, had to obey the ground rules. Mr. Goins' firmness had sent a very clear message that open resistance would not be tolerated.

More Issues

A third issue concerned work orders. Maintenance planners were receiving an increased number of them, the teams making the difference. They did not like the fact that hourly workers were now allowed to write work orders. Maintenance was already overloaded. This just made the situation worse. Another question was, what kind of priority should team-generated work orders receive? Should they be considered equal in importance to, or more important than, those received through normal channels?

The lead team met with representatives of the maintenance planners. Mr. Goins explained that by the time a team-generated work order was written all stakeholders including supervisors and foremen were supposed to have reviewed and approved the project. The immediate response was that in several known instances this had not occurred. The team had written an order without consulting anyone else.

Tom said that this was going to happen. When hourly workers gained a new, attractive privilege they tended to become defensive and to jump the process themselves in order to protect their projects. It was the facilitator's responsibility, in such instances, to make the team retrace its steps and do things properly. When a supervisor suspected that a team was intentionally or unintentionally bypassing stakeholders, he should immediately inform the team's facilitator. This was a learning process for everyone. The hourly workers had to understand that along with their new power they must accept the responsibilities delegated by the ground rules.

Several members of the lead team then spoke in favor of the new, more open management system evolving. Mr. Goins said that the maintenance planners had to be patient and that they should give reasonable priority to team-generated work orders to demonstrate management's support of the team effort. The ground rule concerning response time did not mean team sponsored projects had to be scheduled within one week. It meant that those organizing the actual work needed to get back to the team with a "Yes," a "No, and here's why," or a "Let's talk," within that period.

It was becoming increasingly obvious that Mr. Goins was *the* key actor in this effort. Without his strong backing, too many ways existed to stifle it. Up to this point he had listened carefully to the OP&D consultants' recommendations and said the right things to his work force. If he questioned something, his reasoning had been good and had been presented as a team member seeking the best solution, rather than as a boss. Everyone in the mill was watching him. If he wavered, the OP&D consultants knew that many would back away in order to protect themselves, and that those against the process would grow bolder in their attempts to scuttle it.

Week Eleven

During the week of April 16, the two pulp area teams—operators and maintenance—were led through the start-up exercise. By this time approximately 50 projects had been undertaken by teams in the quality team network. An early concern had been that the workers would focus entirely on their own comfort issues. This did not occur. The majority of projects involved improvements in manufacturing support systems.

With teams whose members were involved directly with production, a majority of their initial priority projects concerned safety—improving communication systems, reorganizing chemical storage, repairing loose guard rails and walkways, putting up warning signs, and making sure people wore their safety glasses. In terms of product improvements, hourly workers, on their own suggestion, began visiting customers to discuss exactly what the customer wanted and how it could be achieved. In terms of management systems, owing mainly to the process ground rules, greater amounts of information were being shared, channels were opening throughout the mill, and, slowly, increased input was being sought from more levels before decisions were made.

A typical project to improve the manufacturing process that concerned a supplier was undertaken by the sheet finishing team in late March. Crews in the sheet finishing function operated a Will Sheeter, which cut, counted, stacked, wrapped, boxed, and palletized sheets of paper from rolls. The team was not satisfied with the cartons being supplied by another IP plant. Because the cartons often carried different shades of coloration, had convex or concave surfaces, or were scored too deeply, the team's problem analysis included the following:

1. The color differences were noticeable and unacceptable to a major customer.
2. The convex surface caused the machine to feed 2 to 3 cartons at a time and to jam.
3. Jamming in the carton machine slowed the sheeter and filled the accumulator belt. At times the jamming caused the machine to be stopped.
4. Concave cartons did not feed into the machine properly and, therefore, had to be hand fed.
5. When running convex and concave cartons, the operator had to tend to that machine only and could not complete other duties.

6. Cartons that were scored too deeply, especially carton bottoms, fell through the carton packing machine before it could set the bottom in place for packing. This caused the machine to jam.

The team worked on remedies for these issues. It then made a presentation to the lead team because this project involved a major outside stakeholder. The lead team's recommendation was that someone from the sheet finishing team contact the supplier and ask that representatives be sent to its next meeting. When this was done the manager of the supplying plant immediately telephoned Mr. Goins to find out what was going on. Mr. Goins said, "It sounds like my people want to talk. Why don't you send someone over?"

On Wednesday of the following week the manager of the supplying plant himself arrived along with two of his reports. They went to Mr. Goins' office. Mr. Goins said, "Why don't you go on downstairs. The team's waiting." So the carton plant manager and his reports sat down with eight hourly workers and one foreman. Doug attended in support of the facilitator, but said little while team members defined the problem, offered their ideas, then worked for three hours with the visitors to generate solutions acceptable to all. By the end of the session the manager of the supplying plant had established a hot-line relationship between the team and his head of production to monitor progress.

Another Typical Project

Another typical project was undertaken by PS&D maintenance. It concerned a loading dock floor. The team's initial concern was tow-motor repairs. Some of the vehicles seemed to be spending most of their time in the shop. Team members began talking about causes of the seemingly excessive damage. Some of the drivers were hotdogging, but this was nothing new. Some needed more training. That could be dealt with, but still did not seem to be the core problem. The team decided to go back through its records and find out what most of the repairs involved. The answer was loose clamp shafts, broken axles, ruptured tires, snapped chains, and out-of-line carriages. Next, team members spent time observing the operation. Almost immediately the major cause of damage became obvious. Sections of the loading dock floor were heavily pocked. When a loaded tow-motor hit one of these holes, it received a severe jolt.

The next step was to do a rough cost-benefit analysis. The team defined the normal yearly cost of the involved repairs. It then estimated the part of that cost attributable to potholes. Next, the team contacted resurfacing companies in the area to get an idea of what covering materials were available and their cost. Once all this information had been gathered, a presentation was made to the lead team because the project involved a fairly large expenditure. The whole dock could be covered with a rubberized cement compound for approximately $80,000. The most used and heavily pocked corridors could be resurfaced for approximately $6500. This was an area of approximately 1600 square feet. During the previous year $16,000 had been spent replacing tow-motor tires alone and $13,000 replacing axles. At least half this amount was estimated to be excessive.

The lead team immediately okayed resurfacing the most heavily traveled area. It asked the sheet finishing team, however, to wait until the purchasing department could check its recommendations concerning materials and suppliers. During this wait the maintenance team developed a more extensive training program for tow-motor operators. Three weeks later the team was told to proceed with its plan as presented. The resurfacing company covered the designated area. The maintenance team was not satisfied with the job and asked the company to redo it. The company balked. The maintenance team went back to the lead team for assistance. Mr. Goins made a phone call. The company redid the job, this time under the close supervision of maintenance team representatives.

The Process Continues to Evolve

As the process itself continued to mature, the issue of quality process training came up. The corporate approach was to train management, facilitators, and team members before forming teams. The OP&D squad had obviously not done that. Rather, it had made training part of more action and results-oriented activities. The most important objective, in terms of this alternative, was to quickly get employees directly involved in systemic improvement efforts. It was to gain access to and effectively begin utilizing their expertise. It was to quickly begin generating highly visible improvements based on this expertise which would help inspire the requisite, overall sense of involvement, ownership, and commitment.

Facilitators and team members had been given, through demonstration, the bare basics in group process and problem-solving skills during the start-up exercise and early sessions. It was assumed that, as they progressed and encountered more sophisticated challenges, the facilitators would identify further process-related needs. The key difference was that the facilitators would be identifying such needs based on their own experience, rather than having the needs defined for them.

During this period the facilitators had formed their own team. They planned to meet every other week to share experiences and to discuss process issues. While the first facilitators had been somewhat hesitant about taking on another responsibility, Sid now had a list of volunteers from a number of staff departments. The realization had come quickly that the position provided excellent, hands-on training in important management skills. It gave those involved access to more people and areas of the mill than they had previously enjoyed. It gave them direct contact with the manufacturing process and its problems. It allowed them to better understand how their own jobs fit into the total operation. Finally, when a team was going well and completing projects, the position gave the young facilitators positive exposure to upper-level management.

The facilitator team's first decision was that each facilitator should handle no more than one team. Otherwise, the load of meetings and support activities interfered too much with normal job responsibilities. Some facilitators complained about the lack of up-front formal training. It was pointed out that they were producing good results without it. When they wanted more training, however, all they had to do was design the session, and it would be theirs.

Another issue brought up by the lead team during this period was how best to alert the rest of the mill work force to team activities. Bulletin boards were hung in each area with notes released by the teams on them. The mill communicator began meeting regularly with Sid and sitting in on selected team meetings. He published accounts of the progress of a variety of team projects in the mill newspaper and released special bulletins.

Ongoing attempts to get potential supporters from Corporate to visit the mill failed. The only Corporate-level interest in the process was shown by the Quality Department. It had developed a list of specific areas to be addressed by all mills as a means of improving quality and productivity. These included energy, dirt in the pulp, water reuse, color uniformity, and fiber loss. The Vice President of Quality told Mr. Goins that his hourly, problem-solving team network would provide a perfect vehicle for carrying out this assignment and should be used for it.

Mr. Goins' answer was, "Fine. We'll do it." Instead, however, he set up task forces independent from the teams but including some of their members to address the issues, honoring the process principle that no one should be allowed to tell teams what to work on, that commitment to the process would come from their sense of project ownership.

During its April 4 meeting the lead team began working on an employee security statement. Two weeks later this statement was distributed to each member of the work force. It said, briefly, that in order for the mill to remain competitive jobs had to change as technology changed. This, however, did not mean that employees would be laid off. Layoffs had not happened as a result of past quality process-driven changes and would not happen in the future. The mill valued its employees and their expertise and did not want to lose either. While lifetime employment could not be guaranteed, employees should be assured that management would do everything within its power to retain them. The distribution of the statement was followed by question and answer sessions for each shift during which Mr. Goins himself fielded most of the questions.

The lead team had also finished its idealization of what was now called the distribution function. It included input from all critical stakeholders. The mission statement for this function now read, "On-time, nondamaged deliveries of exactly what is ordered by the customer." The list of ideal function characteristics had been expanded (see Table 9.1). Design details had been worked out to bring many of these characteristics to life. Implementation steps were being framed and scheduled.

Table 9.1 Final List of Ideal Characteristics for the Product Distribution Function

1. Distribution system handled as a business with suppliers and customers.
2. First step a quality inspection/rejection.
3. Minimal handling.
4. Most live loading.
5. Few scheduling changes.
6. Centralized control.
7. Computerized control.
8. Computerized control of carrier pool and docking facilities.
9. Tow-motor operators needing less supervision.
10. Low-cost carrier used.
11. Trucks loaded from dock closest to paper location.
12. Loading and bracing of trucks and rail cars done according to customer specifications.
13. Minimal reloading.
14. Final inspection/verification.
15. Turn-around time of 24 hours for carriers.

Bump in the Road

By this time the mechanics of the hourly team building effort were fairly sound. Process momentum was building. What remained was to gradually start the rest of the teams. The A/B and C/D roll furnishing teams had covered all their individual issues and decided to form one combined team to work on issues that affected everyone. Power operator and power maintenance teams were brought up during the last week of April. Pulp, woodyard, general maintenance, and technical services were scheduled to be organized in May. The mill-wide network of 18 hourly teams was almost complete (see Table 9.2).

The facilitator team sent a request to Corporate Quality for training in facilitation skills. Two trainers arrived and were impressed with the level of understanding the mill facilitators had already achieved. One of them stayed to become more familiar with the team building process.

Not all of the teams were doing well. Some had difficulty getting members to follow through on action steps. Some had trouble getting members to attend, especially when they were off-shift. A pattern was eventually discernable. Each team developed a hard core of four or five members who attended most of the meetings and did most of the work on projects. Others came and went in cycles. A team would be well-attended and extremely active for several months, then would hit a lull, then would revitalize itself, usually around a

Table 9.2 List of Hourly Teams in Completed Quality Improvement Process Network

PS&D
1. Customer service
2. Roll finishing
3. Sheet finishing
4. Maintenance

Paper
5. Machines 1 and 2
6. Machines 3 and 4
7. Maintenance

Pulp
8. Operators
9. Maintenance

Power
10. Operators
11. Maintenance

Technical
12. Operators
13. Engineers

Administration
14. Controllers
15. Human Resources

General Maintenance
16. Operators

Woodyard
17. Operators
18. Maintenance

new project. While the ground rule concerning attendance was necessary to set the tone, it was obviously not enforceable. The process had to sell itself to employees which, in general, it was doing.

Supervisors and foremen had other, ongoing concerns about the process. One continued to be the work order system. They remained dissatisfied with the way team-generated work orders were being handled. In early May representatives of the supervisors and foremen met with the facilitators and drew up a list of issues (see Table 9.3). When this list had been completed, the combined group designed an approach to handling team-generated work orders that was acceptable to all (see Table 9.4).

Table 9.3 Work Order System Issues

1. How far should facilitators go in weeding out work orders, especially concerning small issues that can be dealt with by the operators themselves?
2. How much maintenance work can production workers themselves do? (This is a union contract issue.)
3. How can area teams keep track of work orders once they are submitted?
4. How do we overcome the team attitude that if their work orders are not addressed, a boss should be fired?
5. How do we overcome the foremen's fear that if they do not address team work orders they will be fired?
6. Is whatever comes out of teams automatically valid?
7. How do we prioritize team work orders?
8. How do we know how many are coming from a team?
9. The system is lopsided. One maintenance area ends up doing all the work.
10. Teams and area foreman have to coordinate better.
11. Production supervisors need to weed out duplications.
12. Foremen need the authority to do some maintenance scheduling.
13. A feedback process needs to be put into place so that teams can learn why supervisors say yes or no to work orders. Ground rule number 12 has to be enforced.
14. What happens when teams will not accept a no response?
15. There needs to be a better system for getting work orders to production supervisors.
16. There needs to be more quality control in the filling out the work orders so that they are understandable.
17. Teams should be forced to justify work orders. Ground rule number 11 has to be enforced.
18. Is a one-week response time to work orders feasible?
19. Will team work orders receive the same or more priority than those already in the system?
20. How many work orders is a team allowed to have in the system at once?

Table 9.4 Improved System for Handling Team-Generated Work Orders

1. Work order priorities set by production supervisor based on project merit, but with special attention paid to those submitted by hourly teams.
2. Facilitators responsible for educating teams so that they understand how the work order system works, how priorities are set, and what the maintenance work load is.
   - Team writes work orders at conclusion of each meeting and puts the team's priority for the project on the work order.
   - Facilitator signs the work order.
   - Facilitator gives work order to production supervisor unless decided otherwise by production supervisor and facilitator.
   - Production supervisor responds to team facilitator on decision concerning work order during week following submission. The response may be:
     a. Yes, it is a good idea and scheduled for . . .
     b. No, I do not think it is a good idea and here is why . . .
     c. We need to talk about this . . .
   - Facilitator responsible for keeping track of work order status until work is completed to satisfaction of team.
   - If the response to a work order is no and the team disagrees with the production supervisor's rationale, the facilitator presents the team's objections to the production supervisor. If the production supervisor continues to disagree, he or she meets personally with the team.

Back in New York, owing mainly to Ron's efforts and the report delivered by the QIP Department facilitator skills trainer following his visit to the mill, the Vice President of Quality agreed to a presentation by the OP&D consultants. It was held in late May. It was attended by everyone in the Quality Department and lasted an entire afternoon. A great number of questions were asked. Corporate Quality seemed to finally be realizing that the bottoms-up team building approach being developed by OP&D complemented rather than competed with its top-down efforts. Further discussions on how best to integrate the two pieces were planned.

But before they could occur, in early June, the OP&D Department was abruptly disbanded. IP was top-heavy compared to other companies in the pulp and paper industry. One of the CEO's objectives was to correct this situation. He let it be known that all staff divisions had to be reduced in size. At the decision of the head of Human Resources, one of the immediate victims was the OP&D unit. Everyone left the company except Doug, who found a job in another area. The Louisiana Mill protested, but to no avail. OP&D had fallen victim to corporate politics.

Trying to Keep It Going

Back at the mill, the team building process continued, now under Sid's direction. A critical mass had, fortunately, been achieved. Enough employees on all levels had become convinced of the new management style's value that the process had gained a momentum of its own. It no longer needed pushing and prodding, only guidance. The facilitator network was, by this time, mature enough to provide that guidance. The mill Manager of Human Resources had also become more aggressive in his role as protector of and spokesman for the teams.

In New York, the Vice President of Quality was shifted to another position in the company. Most of his original staff also left the department during the next year. Corporate strategy was to rotate people through quality, to use it as a training grounds. The new Vice President of Quality was another engineer with no background concerning quality improvement in its broader sense. He was, in effect, starting almost from scratch and paid no attention to what was happening in Louisiana.

Excerpts from facilitator meeting notes show what happened at the mill during the next months.

September 18, 1985

- Facilitators need to survey past team accomplishments to define cost savings.
- Facilitators should meet with superintendents to keep them aware of team activities. Purchase requisitions and work orders should be signed by superintendents.
- Teams in the woodyard, in human resources, and in administrative functions on will be brought up in the first two weeks of October.
- Not every facilitator is submitting team minutes. They must be prepared and submitted to Mr. Goins' secretary for typing.
- Include two or three sentences in minutes about team successes.
- If your team is suffering attendance problems, see Sid.

October 2, 1985

- Facilitators need to update accomplishment posters on hall walls of meeting facility.
- Possibly combine roll and sheet finishing teams.

- Once they run out of things to work on, teams should meet only meet when there are problems, but at least every other week.
- Talk to teams about members not coming when their shift schedule makes it difficult (i.e. meeting at 6 A.M. when they work the evening shift). Is there a reasonable way to deal with this situation?
- Request that all safety items be discussed with foremen before they are brought to the teams.
- Only bring to QIP team problems that the operating system has failed to solve.

October 16, 1985

- Attendance appears to be picking up a little.
- Cost management is going well for the mill, but there are still many areas where costs can be controlled and reduced. Ideas should be reported to facilitators by team members and by the facilitators to Sid and to the mill Manager of Human Resources.

October 30, 1985

- The woodyard operators and woodyard maintenance teams have been brought up. Enthusiasm level is very high.
- Still having attendance problems, especially with teams whose facilitators are not attending our facilitator meetings.
- The process has been encouraged by union negotiations. Union representatives said that the hourly teams are addressing a lot of good problems and are helping reduce the laundry list that needs to be covered.
- The customer service team plans to move three customer service staff physically from their current office location out into the warehouse.
- The Mill Communicator fact sheet has developed a form on which facilitators can submit QIP team news items.

January 8, 1986

- The need to rotate team members was discussed.
- Facilitator meetings will be the first Wednesday of each month.

February 5, 1986

- The process is dragging. What can we do? We came up with these ideas:
  a. Training and direction for facilitators.
  b. Facilitators become firmer in providing team leadership.
  c. Have management speak about QIP at next supervisor's meeting; a lot of supervisors are still not on board.
- Sid redistributed ground rules and we discussed how to better follow them.
- We need overall direction or a philosophy. Sid suggested each facilitator should bring suggestions of what our direction/philosophy should be to the next meeting.

February 19, 1986

- Facilitators and supervisors sat down together to discuss the process and how it can be improved. Sid passed out the ground rules and led a discussion of them.
- One supervisor spoke of the teams' need for management guidance. He felt teams were not getting it from facilitators. He said a team project should be judged on the following criteria:
  a. Has it made money?
  b. Has it solved a problem?
  c. Has it made the place run better?
- A supervisor suggested that teams continue to come up with their own problems, but that ideas for problems to be addressed by teams should also come from managers.
- A supervisor suggested that the team network be reorganized, that teams be made cross-functional and broken down according to shift. He also thought the teams should meet less often. He said that both these changes would be beneficial to foremen.
- A supervisor recommended that facilitators instruct teams to work on mainly cost saving projects.

The process at the Louisiana Mill was still producing results. Obviously, however, there were storm clouds building on the horizon.

Topics for Discussion

1. Discuss the ways in which the mill manager used his power to keep the process on track without taking it over.
2. List possible ways to deal with managers who refuse to cooperate.
3. Compare ways your organization keeps employees informed with those used at the Lousiana Mill.
4. Identify ways that the corporation's negative attitude toward the Louisiana Mill could have been avoided.
5. Discuss why the process began to stall after the OP&D team left.

10 Back to Finish up

After Reading This Chapter You Should Know

- Why the concept of critical mass is important.
- How the facilitator network evolves in the systems approach.
- Why it is important to start building management teams at the same time that hourly teams are built.
- The difference between management-level design teams and hourly-level problem-solving teams.
- How the lead team's role changed at the Louisiana Mill and why.

Stopping the Slippage

During the spring of 1986, Doug, in his new role with the White Papers Division, visited the Louisiana Mill. Mr. Goins told him that while the process was still alive, it was struggling. People were confused as to how they should proceed from this point. Many of the managers were still not on board. He asked if Doug could help. Doug recommended that he try to bring at least Tom, who had returned to academia, back in as a consultant. Mr. Goins agreed. He gained the support of his superior, the Vice President of Operations for IP, but the Vice President of Human Resources and the current Vice President of Quality Improvement resisted. The CEO finally made the decision, and Mr. Goins called Tom.

Almost exactly one year had passed since the OP&D consultants had left the mill. The hourly network had been completed. Sid had brought up the rest of the teams. Fortunately, the suggestion made by supervisors to reorganize the hourly teams cross-functionally along shift lines rather than solely according to function had not been followed. Also, the suggestion to combine maintenance and operators on teams had not been followed. Maintenance had balked. Finally, and perhaps most important, the idea of putting supervisors and superintendents on the hourly teams had not been accepted. Such a change would have been entirely antithetical to the systems approach and would have been extremely difficult to deal with at this point.

Upon his return, after saying hello to Sid and to Mr. Goins, Tom walked through the mill on his own, talking with hourly workers. Somewhat to his surprise he found the enthusiasm level nearly as high as it had been when he left. A great number of improvements had been made. Many more were in the works. He also found, however, a high level of frustration resulting largely from three issues:

1. Middle management's continuing lack of support: Some managers were now resisting the process openly and vocally.
2. A lack of recognition for all that had been accomplished by teams: Hourly workers wondered if upper-level management and the corporation actually knew or cared about the improvements they had made.
3. The fact that the lead teams had stopped meeting long ago.

The facilitator network was relatively healthy. Most of the facilitators were now veterans with at least one year's experience. They communicated well with each other and had been receiving training organized by Sid. They had also begun developing their own training exercises. One home-grown example of such an exercise had involved identifying the different personality types found on teams and discussing ways to deal with them. Four categories of team member characteristics had evolved—hoggers, boggers, froggers, and loggers. The traits of hoggers who never stopped talking were obvious. Boggers tended to get buried in the details of projects and not to want to move on. Froggers were team members who kept jumping around, constantly introducing new projects or ideas in the middle of a solution exercise. The challenge was to keep them focused. Loggers were those who rarely spoke and saw the meetings mainly as a rest period.

Immediate Issues

Two decisions came out of Tom's initial meeting with Sid and the facilitators. One was to give some of the most experienced facilitators the added title of area coordinators. Their new responsibilities would include troubleshooting and coordinating the efforts of all the teams in one area, say PS&D, paper, or pulp. This arrangement would take some of the load off Sid and give the veteran facilitators a new challenge.

The second decision was that the facilitator network should idealize its own operation. It should define a mission, identify characteristics, then redesign its responsibilities if need be. The exercise would include a review of the ground rules first introduced by the OP&D consultants. Perhaps, based on experience, these too needed revision.

It was obvious that the facilitator network was now ready to assume process ownership, which had been a long-term objective from the start. The idealization exercise would be an excellent vehicle for encouraging the involved transfer. It was begun immediately and was met with a good deal of enthusiasm.

The supervisor issue was harder to deal with. Most superintendents, supervisors, and foremen felt they had been victimized by the process. Foremen and supervisors had been forced to watch their traditional powers erode as hourly workers, with the support of top management, took on more responsibility and authority. Also, Corporate was again encouraging the mills to collapse management layers in order to improve communication.

In response to this encouragement as well as to the functional breakdown upon which the team building process was based, Mr. Goins had restructured his management hierarchy. Originally, one position, Superintendent of Operations, had controlled the entire pulp and paper making process, from the woodyard where logs were unloaded to PS&D where finished rolls and sheet were shipped. Mr. Goins had broken this down. He had appointed a Superintendent of Finished Products, which included PS&D and papermaking, and a Superintendent of Pulp, which included pulp making and the woodyard. He had also upgraded the Superintendent of Power to a lead team-level position.

At lower levels in the hierarchy foremen had taken over full responsibility for the actual manufacturing process in their areas. Supervisors had been renamed "project supervisors" and were now responsible for leading efforts to improve manufacturing systems, rather than simply running them. The supervisors did not understand their new role. They felt they were being shoved aside. Their scapegoat was the team building process which they said had precipitated all these unwelcome changes. It was obvious that a way had to be found to incorporate the supervisors into the effort, one which clarified their new responsibilities and made these acceptable.

Tom met with the lead team to discuss the best way of bringing the supervisors on board. He suggested that a network of managerial teams be started at this point to complement the hourly problem-solving teams. The management-level teams should be composed of supervisors and could be called design teams. Someone asked what the difference between hourly problem-solving team projects and design team projects was. Tom answered that while problem-solving teams worked on smaller, isolated changes—safety repairs, manufacturing process adjustments, equipment purchases, environmental improvements such as the installation of fans—design team efforts would be broader in scope and would focus on entire systems such as the shift system, the roll reject system, and the purchasing system.

Someone asked why supervisors could not simply be added to the problem-solving teams. Many of these teams had moved past problems and were now working on design issues. Sid and the mill Manager of Human Resources stepped in to firmly oppose this suggestion. They argued that the design teams must be kept separate from the problem-solving teams so that supervisors could gain the same sense of both team and project ownership that the hourly teams now enjoyed.

Middle-Level Management Feeling Pain

The next task was to convince the project supervisors of the legitimacy of their new role. Tom met first with those from PS&D. The department superintendent also attended at Tom's request. The meeting turned immediately, as was expected, into a "Let's get the frustration out" session. Tom started by apologizing for not involving supervisors in the process earlier and by explaining their new role as defined by the lead team. The following conversation ensued.

Supervisor: "Why are the hourly teams making these improvements?"

Tom: "To improve productivity."

Supervisor: "Where you been, boy? They make them to make jobs easier. That's all these workers care about. They aren't as good as they used to be. They're dumb."

Department Superintendent: "Don't we make them dumb by making all their decisions for them?"

Supervisor: "I resent your insinuation that we don't ask our workers anything."

Department Head: I didn't say that. Now don't get defensive, Carl."

Supervisor: I'm defensive because whenever I speak my mind you go after me. I'll just stay quiet."

Second Supervisor: "How's this different from our five year plan?"

Tom: "You're in charge. Everyone here has had ideas at one time or another on how to improve the operation. You can start right now. You don't have to wait five years."

Third Supervisor: The technology unit's upset because so many engineers are being used as facilitators. They're not getting their work done."

Tom: "They spend one hour or less a week in team meetings. About half of the teams meet every other week now. The facilitator network meetings are held during lunch."

Fourth Supervisor: "You'll never change the way they do things in New York and if you don't change that, nothing much is going to change here."

Tom: "If you come up with a better way of doing things and prove it's a better way, New York will change. It won't make sense not to. What happens at most mills is that people just bitch, or one person comes up with an idea but everyone else is too busy to get behind it. Here's your chance to generate well-thought out improvements and to get the whole team, the whole mill behind them."

After the session Tom discussed with Mr. Goins the fact that gaining the support of these men was not going to be easy. Those running the process had to keep hammering away, stressing what the supervisors themselves stood to gain from the change.

Getting the Managerial Teams Going

By the week of July 22, the facilitators had agreed on a mission statement for their function and had started working on the function's ideal characteristics. Their major focus was training. What types of training did they need? What was the best possible source? When did they want it? Should they simply take what corporate had to offer, look elsewhere, or design more of their own?

Another meeting open to all supervisory-level employees was held. Doug attended this one, having come down from Corporate. Probably because Mr. Goins was away from the mill, some of the supervisors let their frustration flow even more freely.

"This bleeping process is useless. It's a waste of employee's time."

"There's no connection between the team effort and the changes that have occurred. They would have happened without the teams."

"We made plenty of improvements before the teams were formed and will continue to make them after the teams are gone."

"Goins hasn't changed. He'll never change. We don't have any more freedom than before."

"Just another New York project being shoved down our throats."

"This is supposed to be participative, but we're being forced to do it without any say."

"Things were going great without it."

Doug reminded them that not New York but Mr. Goins and his direct reports had brought Tom back in. Tom added that everyone knew that improvements were being made long before the QIP team network was built. All this approach pretended to do was to better organize and integrate employee problem-solving and design efforts and to get more people involved so that more could be accomplished.

"You know what hourly workers do when they get involved in making decisions they don't know enough to make? They screw things up and waste our time with a lot of stupid questions."

After about 1½ hours of listening to complaints and accusations, when the supervisors had begun to calm down a little, Tom asked if there were not some changes the supervisors would make if they had the opportunity. One of the most vocal and negative attendants said, "Sure, just give me the money. There's plenty of things I want to do."

Tom said: "Name something you'd like to change."

"Communication among supervisors."

"Let's work on it right now."

The supervisor scowled. "We are working on it. We've been working on it for years. There's no way we're going to get what we need."

"Just about every team we've brought up has said the same thing. What's the first communications-related change you'd make if you had the chance?"

"The supervisors from finished products, paper, and pulp should meet daily."

"With or without their superintendents?"

"Without."

"No need for input from the lead team?"

"One superintendent could attend."

"Does anyone else need access to the information that comes out of these meetings?"

"Engineering would need it. The scheduling of their work should be affected by decisions made in the meeting."

"How would you get the information to them?"

"Distribute meeting notes daily to everyone interested."

"When should the meetings be?"

"Every morning at 6:30 or 7:00 a.m. so the superintendent invited can report what goes on to the production meeting with Mr. Goins at 8:00 a.m."

"All right. So there's a project you think is important. Write it up, refine it, and present it to the lead team because it affects management policy and because the lead team's a stakeholder."

"Goins would never let it happen."

"So that means you'll just keep doing things the old way and bitching about it. That's the same thing a lot of hourly workers said. 'The supervisors will never let this happen.' But it is happening. The point is that you can either sit on the sidelines and watch your life go by or get involved and fight for the changes you think are important."

"You're a teacher, aren't you? Those that can't do, teach."

"A major difference is that now it's not just one of you going to Mr. Goins with his hat in his hand, it's a team representing all the supervisors going to the lead team."

"What about our jobs? Aren't they planning to get rid of us?"

"There's a corporate move on to collapse layers as I'm sure you all know. You can fight it and lose, or go with it and find new roles. What the process is doing now is offering you a new role, one the lead team thinks is going to be much more interesting and challenging than the old one. But to be effective in the new role you have to know more about what's planned for the mill. Doug has come down from New York to present projections of the mill's future if you're interested. It's apparently the first time this has been done in any mill, and it's being done here because the ground rules ensure your access to any information that is considered valuable to your efforts."

Doug's presentation lasted approximately 45 minutes and was followed by another 45 minutes of questions. By the end of the session it had become obvious that while the process still had few supporters in the crowd, most of the supervisors were willing to give it a chance.

Design Teams on a Role

During the following two months 23 design teams composed of supervisors were formed across the mill to complement the 17 hourly problem-solving teams that had resulted from a series of team mergers. The design teams were facilitated by trained problem-solving team facilitators chosen and moved by Sid. New recruits took over facilitation of the problem-solving teams. The start-up exercise for design teams took 1½ hours; the modified idealized design technique was used. An explanation of the team building process and its objectives was not necessary. Everyone was familiar with it by this time.

For each area represented, Tom and Sid had team members identify:

1. Key stakeholders to that area.
2. Technical systems important to the area operation when a line unit was involved.
3. Administrative procedures important to the area operation when a staff unit was involved.
4. Key management systems that impacted the area's productivity.

Teams then picked the technical system, administrative procedure, or management system they thought needed the most improvement. They generated a list of questions they wanted to ask concerning that system/procedure. They developed a list of characteristics the system/procedure should have ideally. They then began defining action steps necessary to either discover the answer to their questions or to redesign system/procedure parts so that they took on the desired characteristic. Finally, they identified stakeholders whose input was required before changes were implemented and discussed the design with them. The results of the start-up exercise for the pulp operation team can be found in Table 10.1.

The initial projects embarked on by these teams included the redesign of the standard cost system, the wood handling system, the stock prep system, the mail system, the roll wrapping system, the bleaching system, the internal communication system, the digester system, the steam allocation system, and the time reporting system.

Best Role for the Lead Team

On August 5th, Tom began working with the lead team again. Lead team meetings had been combined with morning production meetings. QIP team issues had not faired well against operational issues in the ensuing competition for time. The first step, therefore, was to encourage the lead team to reintroduce a weekly session dedicated solely to improvement and to quality process issues.

Table 10.1 Pulp Area Design Team Start-Up Exercise Results

1. External stakeholder:

Paper mill	Power house (WTP)
Woodyard	Evaporators (hot water)
Caustic plant	Railroad
Bleach plant	Truckers (suppliers)
Engineering (process)	Environmental services (clarifier-reclaimer)
Unions	Purchasing
Administrative services	

2. Key technical systems:

Digester	Liquor recovery
Chip silo	Storage
Wet room	Testing
Warm water system	Instrumentation
Turpentine	Communications
Condensate	

3. Key management systems:

QIP	Shift system
Union	Motivation
MTS	Communications
Safety	

4. System most in need of improvement is digester. Important questions concerning it include:
   1. What causes it to hammer (TC vibration)?
   2. Why are P numbers so erratic?
   3. What causes internal scale build-up?
   4. Concerning daily production:
      - How much are we making?
      - How much should we make?
      - What is it designed to make?
   5. How should it be controlled?
   6. Why variables cannot be consistently measured.
   7. Why we cannot have good standard chips.
   8. Why we cannot get maintenance process training.
   9. Why not buy chips?

5. Characteristics digester system should ideally include:
   1. Uniform chips.
   2. Uniform liquor.
   3. Good level control.
   4. Wash liquor quality.
   5. Temperature control for steam.
   6. Design capacity greater than daily needs.
   7. Do continual training without overtime.

(continued)

8. Have skilled experienced digester operators.
9. Good communication with process suppliers (liquor, chips, etc.).
10. Train engineering with maintenance and operators.

6. Action steps to be completed before next meeting:
   1. Check with Hammermill process groups for their method of measuring white liquor.
   2. Check with woodyard to use number 5 silo as own-made chip silo. This was done before, but is no longer standard operating procedure.

The first such session went well. The atmosphere had changed greatly. Mr. Goins was there, but no longer dominated. His reports were relaxed with each other and with him. All of them, with one exception, began participating immediately. Now that supervisor-level teams were redesigning departmental production systems on a mill-wide basis, the decision was made to focus on three critical macro-issues that affected everyone. The first was the mill's relationship with outside stakeholders such as Woodlands and the Corporate Customer Service staff. The team felt that several of these relationships needed realignment.

The second issue was training. The corporation had made major changes in its training program over the last several years. Not all of them had been acceptable to employees for a variety of reasons. Several hourly teams had already attempted to deal with local training issues, but their efforts had been piecemeal. The lead team decided to develop a comprehensive and realistic model of its own for employee training that would incorporate technical training as well as training in statistical process and quality control for which they felt the work force was now ready.

The third issue involved discovering ways to increase operational effectiveness. The lead team decided to explore the possibility of turning key mill areas—the woodyard, pulp, paper, power, PS&D, maintenance, and so on—into a continuum of profit centers. The controller had to first come up with a system for defining costs and profits, then define what needed to be measured. After this the engineers needed to develop standards and a means of measurement.

Two lead team members were given responsibility for developing projects around each of these issues. The team also defined its own mini-set of ground rules to govern its output. These were:

1. Everyone affected must have the opportunity for input.
2. Designs must be technologically feasible and financially appropriate.
3. Results must promote an ethical business philosophy.

4. Results must promote employee security.

As time passed, however, very little happened. Everyone continued to participate. Action steps, however, were not completed and the projects eventually bogged down. It became obvious that at this level to work on projects outside team meeting hours was an unwise move. Too much was going on operationally. The mill was preparing for union contract negotiations. Also, Corporate headquarters was moving from New York City to Memphis, TN. A major shakeup was occurring owing to the expected loss of headquarters personnel resulting from this move.

At that point Sid suggested the lead team's focus be shifted away from design issues and concentrated, at least in the short term, on its third responsibility—reviewing, contributing to, and helping coordinate the efforts of supervisory- and hourly-level teams. Such a shift would relieve lead team members of unwanted outside work and, at the same time, increase the team's visibility.

The facilitator network had, by this time, designed its ideal self. Its mission was:

> "To help the Louisiana Mill be the best by managing the development and implementation of improvements originated by QIP teams."

The ideal systems characteristics defined as necessary to the realization of this mission were:

1. Management support.
2. Adequate training.
3. Dedicated facilitators.
4. Career advantages.
5. Good understanding/support from middle management.
6. Good communication among facilitators.
7. Well-defined goals.
8. An apprenticeship program for facilitator trainees.
9. The ability to coordinate hourly with management-level goals.

The facilitators began working on the actualization of their number two characteristic, adequate training. They defined the types of training desired, then began exploring sources. Within two weeks a course in problem-solving techniques had been scheduled for all facilitators. Those who had not yet taken the corporate leadership/communications skills course were signed up to do so. A new, refresher course in this same area was being designed. A system to circulate relevant articles through the facilitator network was being put into place. Finally, the group was developing an apprenticeship program for newly recruited facilitators.

Final Issues

By this time the team network was complete and running smoothly. Sid and his lead facilitators (area coordinators) were making process decisions. Tom was simply advising. The Corporate CEO had stated bluntly in a recent speech that people who wanted to stay with International Paper and to progress must actively support QIP. This sentiment was repeated by Mr. Goins to his supervisors, a diminishing number of whom were still resisting the change in their role.

During Tom's ongoing conversations with the supervisors, some of them said that they were willing to push decision-making authority downward into the hourly ranks, but that when something went wrong they were the ones who got blamed, so they were afraid to take the risk. Tom brought this issue to the lead team, which decided that more stress should be put on the positive. Employees should not be afraid to make mistakes when they were trying hard, so long as they learned from them. A differentiation was made by one lead team member between honest mistakes and those caused by not caring, by sloppiness. Too many of the latter type were still occurring. It was pointed out by another member, however, that when the process was working properly peer pressure became the main form of punishment for goofing off, rather than a bawling out by the boss.

Another issue brought up at the lead team meeting was that several of the managerial design teams were having difficulty understanding whether their assignment was to redesign entire areas at once or whether they could focus on the systems within their areas incrementally. The lead team eventually decided to leave this decision up to the design teams. They could begin with increments. As more and more pieces of the puzzle were completed and had to be fit together, however, the need for an overall area design would become apparent on its own.

Facilitators Take the Lead

The facilitators held their first home-grown work shop on two consecutive days over lunch. Lead team members were invited. The sessions were well attended. The first day 31 facilitators came and the second day 33 came. There were now a total of 41 facilitators. Those who missed were on vacation, taking courses, or in foreman slots where they could not leave the floor.

The workshop was built around the following agenda:

1. Review and discussion of ground rules, which facilitators had revised to better fit the mill's situation.
2. Review and discussion of the facilitator's role (see Table 10.2).
3. Skit portraying a typical team meeting, senior facilitators demonstrating some of the problems that arise and how they might be handled.
4. Participants divide into teams, work on prepared situations, present results, and discuss alternatives.
5. Participants review and clarify steps in team design or redesign exercise.

The agenda also included a conversation with the lead team representatives who attended the session. The questions addressed included:

1. Does a team have any recourse if it does not agree with a supervisor's rejection of a team solution to a problem addressed?
2. What kind of involvement/direction can design teams expect from the lead team?
3. How much time should be spent during work-hours on facilitation of QIP team projects?
4. What is the priority of QIP items versus day-to-day items?
5. Is a seven-day response time feasible? If so, how can the lead team help encourage the required response?

One of the remaining process issues was recognition. A need existed to encourage the generation of increased applause for individual team accomplishments on a mill-wide basis. Several steps were taken. The first was to compile a record of the more than two hundred and fifty projects completed by teams thus far. A booklet containing this record was printed and distributed so that employees could begin to gain some idea of the big picture, of what had been happening in parts of the mill other than their own, of the breadth and volume of activity. A second step was to videotape some of the most rewarding projects for distribution inside the mill and throughout the corporation. Finally, an effort was made to discuss at least one project a week in the mill newspaper.

Table 10.2 Facilitator's Roles

1. Coordinator of pre- and post-meeting logistics:
   - Agenda
   - Minutes
   - Resources
   - Time
   - Location
2. Team trainer:
   - Ground Rules
   - Problem-solving techniques
   - Work order process
   - Patience
   - Presentation skills
3. Referee:
   - Between team members
   - Between team members and outside stakeholders
   - Between team members and outside trainers
4. Coach to provide:
   - Focus
   - Guidance
   - Motivation
5. Protector of:
   - Individuals
   - Ideas
   - Company interests
   - Resources
6. Equalizer to keep balance of :
   - Participation
   - Projects
   - Business/socializing
7. Cheerleader to provide:
   - Praise
   - Encouragement
   - Active support
8. Press Agent to insure team gets:
   - Recognition
9. Interrogator to perhaps most importantly:
   - Ask questions!

In terms of Corporate recognition, while Mr. Goins had been asked to deliver speeches to several groups at the Corporate level, little else had occurred. Only two people had visited from Corporate since the Quality Department trainer had come nearly two years before. The manager of one other mill, Androscoggin, had also sent two people down to see what was going on. No other mills had shown interest.

By the end of 1986 Tom had stopped traveling to the mill. He was now on call. Instead of scheduling visits himself based on what he thought process needs were, he left it up to Sid and Mr. Goins to decide when and if he should come down. Otherwise, they were free to call him anytime. Sid called approximately every other week for several months, then monthly to discuss what was going on. The team network had become a simple but effective three layer affair including the lead team, then the supervisors' design teams, most of which now included foremen and superintendents, and, on the bottom (top?) the hourly problem-solving teams, almost all of which now included one or two foremen.

The IP Corporate Quality Department, in the meantime, had focused on long-range planning as *the* answer. It was still paying no attention to what was going on in Louisiana. One member of the lead team summed the situation up nicely by saying that the rest of IP was not paying attention because it just was not ready yet. These things took time.

Topics for Discussion

1. Are facilitators capable of directing a quality improvement process?
2. How did your organization get managers involved and did its approach generate commitment?
3. Why is there often a need for management teams defined by level as well as by function?
4. Are the steps your organization has taken to incorporate strategic planning with its quality improvement process effective from a systems perspective?
5. How do you know when a quality improvement process can sustain itself without outside help?

11 A Very Different Situation

After Reading This Chapter You Should Know

- Whether a quality improvement process can be successfully introduced when the work force is entirely new.
- How to deal with managers who insist on establishing cross-functional teams.
- The tie between safety programs and quality improvement programs.
- How a quality improvement process can lead to a total reorganization of the management system.
- Various forms of low-level recognition.
- One way of introducing statistical measurement techniques successfully.

Into the Trenches

In February of 1987, Tom was invited to visit IP's Androscoggin Mill in Jay, ME. This mill, another one of the largest in the system, housed five paper machines. Approximately 1200 hourly and 185 salaried employees ran the operation. The mill produced a wide range of paper products, as well as flash dried and raw pulp.

Owing at least partially to the diversity of operations, the mill had for a long while not generated the profits it was capable of generating. Another reason for its sub-par performance was a history of strained management-union relations. An increasingly adversarial stance had been developing until, when Tom arrived, a strike was a serious possibility if not a probability.

Androscoggin differed from the Louisiana Mill in that it had a Quality Improvement Department headed by a manager, Belinda Bothwick, who reported directly to the mill manager, Newland Lesco. Androscoggin had been involved in efforts to improve product quality for several years now. Most recently, an attempt to introduce statistical quality control (SQC) techniques had been made. It had failed owing to the inability to generate the necessary level of employee commitment. Also, earlier, a large number of quality circles had been formed. This tactic had also foundered, owing mainly to a lack of necessary integration.

Tom spoke to a gathering of Mr. Lesco's direct reports, outlining the team building paradigm introduced at the Louisiana Mill and the logic behind it. After questions were answered the managers voted, by a narrow margin, to hire him. However, despite Tom's argument that the process would help improve management-labor relations and help provide alternatives to a strike, the managers also decided to wait until after the union contract had been negotiated to begin.

In June of 1987, two of the three mill unions, the UPIU and the International Brotherhood of Firemen and Oilers (IBFO) rejected the contract offered by the company and struck. The office workers union did not. The corporation, in reaction, began almost immediately replacing the strikers. Most of the approximately 1200 replacement workers hired at this point had no background in pulp or paper making. The situation was made worse by the fact that quite a bit of damage had been left behind by the departing strikers.

Time to Get Started

In November of 1987, Tom was asked to return to the mill and start the team building effort. Everyone there was tired. Shifts of 12 and 16 hours were not uncommon. Some managers were still on the floor running the machines, as they had been since the early days of the strike. The mill was deeply involved in crisis management. The staff was trying to produce paper, train the new hourly work force, repair the damage left behind, and upgrade process technology all at once.

Many senior managers were skeptical about beginning a team building effort at this point. Too much was already going on. Also, the new employees had not been around long enough to understand the manufacturing processes they worked with, much less improve them. Belinda's counterargument was that this, indeed, was the ideal time to change the management culture. Bad habits had not formed yet in the new work force. Management-labor relations were still maleable. The new workers were enthusiastic rather than cynical and were eager to make a positive impression.

It was finally agreed that teams would be formed only in PS&D. If these produced anything of value, the process would be spread throughout the mill. If not, it would be put on hold again. Belinda and Tom, taking as a given that the process would go mill-wide, broke the entire operation down by function. The optimal number of workers for a team to represent was between 40 and 60, though Customer Service included only 13 and other functions close to 80.

An important event occurred at this point. The mill manager, Mr. Lesco, who had brought the process in and supported it, left Androscoggin to take a position at Corporate headquarters. The new mill manager, Jim Thompson, was by reputation a strong advocate of employee involvement. Situational demands, however, consumed most of his time and frequently took him off-site so that he was unable to provide the constant and extremely visible support and pressure Mr. Goins had provided at the Louisiana Mill. He answered all calls for help and followed all process-related suggestions, but he was simply too new and too busy to function as a major resource.

Belinda and Tom turned to the department superintendents for support. At Androscoggin there was one for each major area—PS&D, paper, pulp, power and maintenance. Fortunately, the superintendent of PS&D became a strong advocate when the teams in his area started producing positive results almost immediately. Most of the other department superintendents followed suit, but not all. When the decision was made to expand the team network beyond PS&D serious resistance surfaced immediately in the paper machine area. The department superintendent there was an old-timer, a hands-on type with tremendous technical knowledge. When a machine breakdown occurred he appeared, orchestrating the necessary repairs or rolling up his sleeves and making them himself.

Market demands and the corporate reward system continued to place emphasis mainly on the number of rolls, sheets, and so on produced within a given period of time. Quality remained a secondary consideration in most instances. When a paper machine went down, therefore, the objective was to patch it up and get it running again as quickly as possible. To stand back and try to teach new workers what to do made no sense to this old-timer. The participative, team concept also made little sense. It slowed things down. Eventually, this man left the mill and was moved to a corporate staff position as a technical consultant.

The new head of paper, though also a strong leader with a strong technical background, immediately began delegating authority and encouraging his subordinates to do the same. Despite his heavy schedule, he eventually became an hourly team facilitator himself, as did several members of the lead team, making themselves more accessible and responsive to the work force.

The paper-making operation included a number of different functions. It was suggested that putting all of these functions on the same team would improve communication between them. This was tried, but with predictable results. Representatives from the individual functions wanted immediately to address issues unique to their function. It was extremely difficult and often impossible to reach consensus on priorities. The teams ended up undertaking several unrelated projects at the same time.

Eventually, the paper-making team fragmented and had to be reorganized. The reorganization put employees serving the same function for each of the five paper machines on the same team so that they had common interests and could learn from each other. A problem brought up by representatives from machine number 1, for example, might already have been solved by machine number 3.

Threats and Logistics

The new employees adapted quickly to the team approach. They were willing to attack just about any problem. Very little was accepted as a given. At the same time, they began using the team as an employee-controlled learning vehicle. Management quickly came to see the team network as a means of speeding up employee training and as a way to give the work force more authority without losing control.

It was not long, however, before some of the superintendents also began seeing the teams as a vehicle for addressing their own priorities. Because these priorities were often critical to the survival of the mill, it was difficult for Belinda, Tom, and several of the more supportive lead team members to explain that:

1. The major purpose of the team building effort was to generate commitment to improved quality and that commitment came from a sense of ownership. If managers started dictating team projects, it would be business as usual. During their one hour a week meeting the

teams should have complete control over what they worked on, bound only by the ground rules.
2. The teams usually picked projects important to the managers anyway. The difference was that they were doing the picking.
3. Managers could form a task force any time they wanted to address a project. They could enlist any employees they wanted for such a task force.
4. Supervisors, at least, had QIP teams of their own to which they could bring such projects.

Owing to the pressures of reorganizing the mill operation, some of the superintendents had trouble accepting this rationale and continued their efforts to take over the teams. Eventually Mr. Thompson had to intercede and make it absolutely clear that managers could not tell QIP teams what to work on.

Tom trained Belinda immediately to lead the start-up exercise for both problem-solving and design teams. After several months Belinda was given an assistant, Peter Allen, who was also trained. Most team meetings were held in a row of classrooms on a balcony above the PS&D area. The team accomplishment posters were hung in the hallway that connected the office building through which most employees entered the mill.

In comparison with the Louisiana effort, the lead team's self-defined process-related responsibilities were similar but richer. They included:

1. Developing an overall framework of mill objectives into which mill teams could fit their projects.
2. Identifying mill-wide issues on which the lead team or task forces could work.
3. Contributing as a stakeholder to any problem-solving or design team project that involved changes in policy or large expenditures.
4. Setting the example, helping overcome the reticence of middle managers by giving them more decision-making authority in their areas of expertise.
5. Being available to all teams as a resource.
6. Functioning as a reward to teams by listening to select, short presentations on outside projects as defined and scheduled by the head facilitator.

Potential team facilitators volunteered from all areas of the mill. As in Louisiana they were not allowed to facilitate a team representing their own function for fear of them forcing their own priorities and improvement-related ideas on the others. As was eventually done at Louisiana, area coordinators were also appointed for each department to troubleshoot the QIP teams in that department, to help integrate team efforts, and to help identify training needs.

The mill manager's secretary again took initial responsibility for organizing and feeding the process tracking system. She received copies of all team notes that had been released. She developed two master files, one paper, one computerized. Team folders in the paper file were color-coded according to department. These files included:

1. A team project form that was generated to record monthly progress on larger projects and to record the results of smaller projects. This form included a space for a cost-benefit analysis and for the recording of actual savings. A copy of the form filled out for each project was placed in both the paper and the computerized files.
2. A team change notification form that was devised to track changes in team membership. It included a notation on the team meeting time, day, and location. It also included a list of those who were to receive the team's notes.
3. A meeting schedule of QIP teams that was developed to consolidate and keep track of the weekly meeting times and meeting locations of all teams. This was published monthly.
4. A distribution list that was designed for each department listing all the teams in the department, their members, their facilitators, and their computer document numbers. Modifications in the team member lists were based on information gleaned from the team change notification form.

The QIP-Safety Connection

Another relationship requiring attention at Androscoggin was that between the quality improvement effort and the safety improvement effort. IP, like most modern-day corporations, was extremely safety conscious. Also, like most modern-day corporations, however, its approach had several important weaknesses. The typical safety program includes the following elements:

1. Posters and other educational materials.
2. Presentations on relevant safety issues, which act as a reminder of things employees should already know and as a warning concerning new dangers in the work place.
3. Periodic crew meetings to pinpoint local safety issues and sometimes to define solutions.
4. Accident reports, monthly records of lost-time and nonlost-time accidents.
5. A head of safety at the facility level responsible for overseeing all the above.
6. A corporate safety director responsible for coordinating the education program and for reducing accidents on a corporate-wide basis.

Problems with the typical safety program include the following:

1. Lack of ownership by those most important to the program's success.
2. Absence of an appropriate vehicle for instituting safety-related changes in the work place.
3. Dependence on an incentive system that had the potential of actually hampering efforts to make the work place safer.

Concerning the ownership problem, while employees and employee safety committees identify problems, it is usually left up to management to define and to implement the necessary changes. Employees, therefore, do not make the contribution they are capable of making. Members of the work force would be more likely to stress safety if given the power to implement changes they themselves thought necessary or wise.

This approach makes sense for the following reasons. First, in most situations workers know better than managers what the safety problems are because they must deal with them on a continuing basis. They also know what the most appropriate solutions are for the same reason. Second, because there are more workers than managers, if given the necessary authority and access to resources they can deal with more problems in a shorter period of time. Third, foremen and other managers usually have a greater number of responsibilities than workers, safety-related improvements being just one. They also frequently have a different set of priorities, so that safety issues important to workers take a back seat. When workers have ownership a number of steps in the correction process are eliminated and constant monitoring and proper feedback are better assured. As a result, the chances for communications foul-ups, for a work order to get lost or buried, for example, are reduced.

In terms of an effective change vehicle, very few corporations have put into place that which is necessary. What one generally finds is projects competing for attention on all levels. Individual employees or small groups bring safety issues to foremen or supervisors. Foremen and supervisors bring their competing lists to the maintenance department. Maintenance combines these lists with lists from other sources and tries to decide which projects to schedule first. Those involving a production process breakdown and those submitted by top-level management are generally given priority. Safety projects, unless a crisis or lost-time accident is involved, frequently have to wait.

Keeping It Legitimate

In terms of incentives, traditional emphasis has been on reducing the number of accidents, the key word here being number. As a result of this approach, the corporate-level safety director is graded on how effective he or she is at cutting the number of accidents reported, the key word here being reported. Such reductions occur in three ways:

1. Through luck.
2. Through legitimate educational efforts.
3. Through illegitimate maneuvering.

Illegitimate maneuvering involves finding some way to misrepresent reality. One means of doing this is by simply fudging the figures. An extreme example would be to not consider an accident reportable unless the victim is carried out on a stretcher. Another not so extreme example would be to give an employee with a banged-up leg the job of monitoring the infirmary waiting room television screen until he or she can return to normal duties so that no lost time need be reported.

A second, more subtle strategy is for the corporate safety director to let it be known that accident rates are expected to decrease and that manager's careers could be adversely affected if they do not. The resultant actions and attitudes of managers force workers to begin stocking their lockers with medical supplies. If an injury is too serious to deal with in the locker room, an excuse is found to go home early or to a hospital, co-workers covering.

A third approach is to hassle the workers who do bring their injuries to the infirmary, to make them fill out long accident reports, to investigate the accident in order to discover the degree of negligence involved, to broadcast the injured worker's name along with an account of the crime.

As a result of management's current emphasis on the numbers, its emphasis on racking up a record total of accident-free days, and its quantitative as opposed to qualitative thrust, safety programs frequently end up alienating the very people whose quality of working life and productivity they are designed to improve. Employees end up becoming the victim, rather than the beneficiary, of the involved efforts.

There are several ways to make safety programs more effective. Perhaps most important, top-level management should stop broadcasting the numbers, at least at the corporate level. This practice puts emphasis in the wrong place. It turns safety into a competitive rather than a cooperative effort. It places unrealistic demands on the corporate-level safety program director, demands that can warp his or her perspective and cause that person, in turn, to place unrealistic demands on unit safety heads.

The corporate director's role should be threefold. First, that person should function as a resource. The director should make sure that all units know about and have access to training materials. Second, that person should facilitate the organization and integration of training efforts throughout the company. Third, the director should make sure that top-level management is kept abreast of developments and that it continues to demonstrate highly visible support for the effort.

The head of safety at the unit level should again act as an organizer and integrator for safety improvement efforts. This person should make sure an appropriate change vehicle exists and that employees have access to all useful resources. This person should also be in charge of mounting an effort to encourage employees to seek help at the infirmary when injured, and should strive to make the involved process as painless as possible. The espoused theory should be that workers do not want to get hurt, and that when they do get hurt the company wants to help, rather than embarrass them. Records of the number and types of accidents which occur at the unit level should be used to pinpoint trouble areas in the operation, rather than to judge individual performance.

Safety, in effect, should no longer be something done to workers, or taught to them by management. Rather, it should become something that the workers teach themselves, oversee, and ensure. It should become part of their everyday activities, their self-defined job responsibilities. Safety and quality improvement are inseparable. When employees are given responsibility for improving the quality of manufacturing processes and the working environment, safety is, almost without exception, their first consideration. Unfortunately, IP did not yet understand this. It ran the safety and quality efforts independently so that duplication and competition for resources frequently occurred.

At Androscoggin safety had to be the number one concern owing to the new, unskilled work force, the number of repairs and modifications being made, and the fact that the striking union had lodged a number of complaints with OSHA, precipitating a formal inquiry. A new Director of Safety joined the Human Resources Department staff at the same time Tom began working with Belinda and eventually launched a safety education and improvement effort complete with area teams. This was unfortunate because at least 40 percent of the initial improvements made by the hourly QIP teams were safety related.

Tom and Belinda suggested as an alternative that the Safety Director focus on education; that task forces governed by the same ground rules as the QIP teams be formed around specific safety projects identified by the lead team and others; and that, rather than forming her own teams, the safety director help facilitate and integrate the safety-oriented projects already being generated by the QIP teams. The new director, however, balked at this strategy. She said that she had to follow the procedures dictated by her boss at Corporate.

In early 1989 safety and QIP were, indeed, put under one Vice President at the corporate level. Little, however, was done to integrate the efforts. Their similarities in terms of both objectives and approach were still not appreciated.

Restructuring the Mill Management System

Belinda had been meeting with groups of managers and with individual managers trying to help them understand and accept their role as it was being redefined by the QIP. During one of these sessions with the mill's area supervisors, participants began identifying systems improvements specific to their level which they wanted to address. Belinda suggested that the group turn itself into a QIP team representing a level of management, rather than a function, and join the QIP team network.

Tom took initial responsibility for bringing up and facilitating this new team. During the start-up exercise the cross-mill supervisor-level team's number one priority quickly became obvious. It wanted a clear definition of supervisor-level responsibilities. It also became clear that the team members wanted their bosses, the departmental superintendents, to step aside and let them run the day-to-day operations. Finally, it became clear that they saw little possibility of this occurring. The old ways of doing things were too firmly entrenched.

Tom eventually convinced them to participate in the design of what they considered to be the ideal situation. Early team meetings were poorly attended. A hard core of approximately ten always showed up and did most of the initial work. As the design began to evolve and make sense, however, enthusiasm built. Eventually 20 or more supervisors began attending.

Within two months the team had designed an ideal management system strawman. The design that team members came up with included an *operations coordinating team* on which area supervisors representing every department sat. This coordinating team met every morning for members to share information, to discuss the previous day's operation, and to coordinate that day's planned activities. The coordinating team, in turn, was to be fed by 13 *area operations teams,* one for each key production function. These would be composed of:

1. All area supervisors working in that function.
2. Representatives from other levels of that function.
3. Representatives from other functions affecting that function.

Membership of the 13 area operations teams was to be broken down into core members and consultants. The core members would attend on a regular basis. The consultants would attend on as as-needed basis. The composition of the groundwood area operations team can be found in Table 11.1. The area operations teams were to take charge of planning and overseeing the daily manufacturing operation. The issues these teams were given responsibility for included:

1. Quality.
2. Safety.
3. Production levels.
4. Work environment improvements.
5. Requirements concerning hourly employees.
6. Costs.
7. Production planning.
8. Maintenance planning.
9. Morale.
10. Training.
11. Scheduling.

Table 11.1 Area Operations Team Composition for Groundwood

Count	Position
Core Members:	
1	Area supervisor
1	Shift supervisor
1	Operator
1	Maintenance foreman
1	Maintenance analyst
1	Equipment reliability foreman
1	Preventive maintenance manager
1	I and E maintenance representative
1	Project engineer (department)
1	Training coordinator (if existing)
10	
Consultants:	
1	IR representative
1	Engineering representative
1	Finance representative
1	Purchasing representative
1	Safety representative
5	

Finally, a representative of the coordinating team would regularly attend the mill managers morning production meeting to report its activities and gain input from top level management. The new primary role of top level management in this scenario, instead of overseeing day-to-day operations, would now be to develop long-term mill objectives for the coordinating team to base its decisions on. In turn, top level management would be required to incorporate priorities defined by the coordinating team into its plan.

Several members of the mill manager's group became deeply concerned about this project, suggesting that it exceeded the charter of the QIP. An attempt was even made at one point to force the area supervisors to drop it. Belinda and Tom went to the mill manager and explained the situation. He responded immediately by attending an area supervisors team meeting and voicing his support for this particular project. Eventually, four of the most active members of the area supervisors team presented the preliminary design to top level management as a major stakeholder for its input. During this exchange most of the confusion and misinformation that had led to the opposition was cleared away.

Next, the area supervisors team began working on project action steps. These included:

1. Put formation of the coordinating team on hold until area operations teams are in place.
2. Work on developing one or two area operations teams. Get them in place and functioning properly before forming any more.
3. Develop area operations teams in areas that already have hourly problem-solving and supervisory design teams in place.
4. Have the department manager and area supervisors identify area operations team core members and consultants.
5. Familiarize all employees in each area with the area operations team's role and responsibilities.
6. Experiment with how often area operations teams should meet, when, and where.

While some of the area operations teams started functioning almost immediately, it quickly became obvious that getting all 13 in place would take a long time. The area supervisors, therefore, decided to go ahead and form the coordinating team. Very quickly the work of this group gained the blessing of top level management, relieving top level managers of day-to-day issues which had consumed a great deal of their time.

Eventually, production meetings were held only three times a week instead of every day, and participants were able to focus on strategic issues during these sessions. At the same time, top level management remained fully informed on coordinating team activities and decisions and retained the right to intercede when it thought intervention necessary. The mill management system which evolved as a result of this project is outlined in Figure 11.1.

Recognition

The area supervisors' team success gave the overall process a substantial boost. By the latter part of 1988 the targeted critical mass of support definitely had been generated. The process had taken on its own momentum. Belinda and the facilitators possessed the understanding and skills necessary to keep it on track. Enough successful projects, over 150 by this time, had been produced by teams to convert most critics into believers and to move the rest to silence.

Recognition became an important issue. Employees were wondering if anyone was really noticing the improvements they were making. The facilitators brainstormed this issue and developed a list of possible awards including:

Figure 11.1 Mill Management System Defined by Area Supervisor's Team

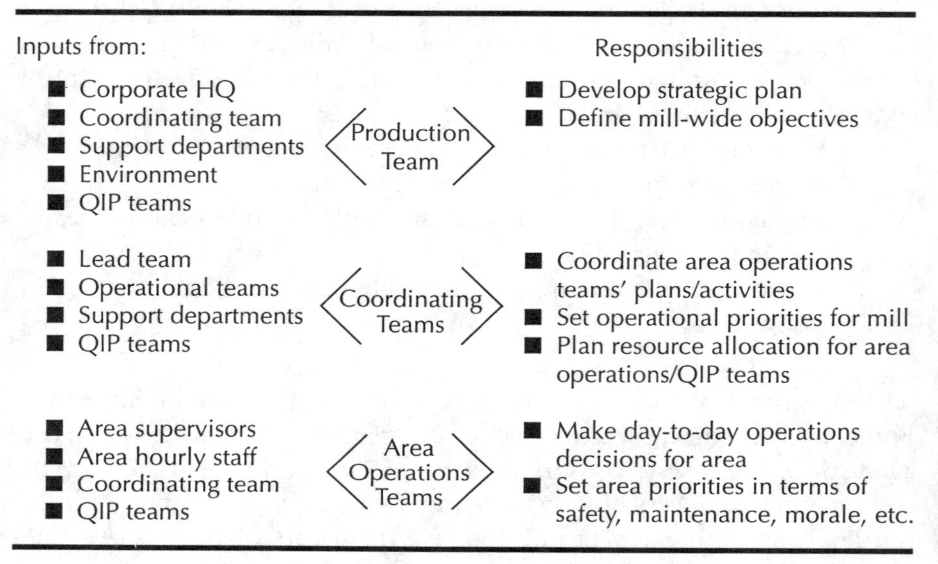

1. Letters of congratulations.
2. Mill manager, department heads, supervisors visiting team meetings.
3. Letters of commendation added to personnel files.
4. On-floor recognition by supervisors.
5. Team T-shirts.
6. Gestures of personal recognition.
7. Appropriate gifts.
8. Opportunities for work exchange between members of different teams.
9. Designated parking spaces for members of teams that generate outstanding improvements.
10. Articles on projects in the mill newspaper.
11. Photos added to team status sheets on hallway leading to the mill.
12. Project progress bulletin boards.
13. Moving award (trophy) for team good project competition.
14. Luncheons for team working on the same project.
15. Meetings occasionally catered with coffee, doughnuts, and so on.

At the end of the year, when teams started celebrating their anniversaries, a luncheon was held for each and attended by the department superintendent and the mill manager whenever possible.

Trying to Get the Foremen on Board

By the beginning of 1989 the only population that the lead team felt had not been incorporated sufficiently into the effort was the foremen. Some sat on hourly teams, but here they were usually unable to address the issues important to themselves. Some participated in department design team meetings. Some sat on task forces. The majority, however, were not part of any team. As a group they were bothered by a number of important issues. One was a persistent rumor that their level of management was eventually going to be eliminated. Another was their changing role, hourly workers taking away increasing amounts of their traditional responsibility. They definitely needed a forum where they could at least discuss these threats.

The foremen worked swing shift. Any hour picked for a meeting of the whole, therefore, would exclude at least one-half of them. Also, there were too many foremen for one meeting. The chance to participate would be too limited. The decision was made to break this group down by shift, one team per shift including representatives from across the mill. The shift teams would be encouraged to address level issues, at least initially, as the area supervisor team had, rather than department or function specific issues.

The initial sessions during which the process was described and discussed and potential projects identified were well attended. The majority of the foremen seemed interested in having their own teams and working on their own projects. When the actual team meetings began, however, very few showed up. Many of those who did were the same foremen already active on hourly teams. Others had been encouraged by their bosses to do so. Projects began, but were rarely followed through. Reasons for the lack of attendance included emergencies, scheduling difficulties, and forgetfulness.

Eventually, the teams were disbanded. The mill manager set up, instead, a weekly information session hosted by himself, which all foremen were invited to attend and during which any topic or problem could be discussed. The turnout was relatively good. Foremen were also encouraged to contribute to the strategic planning exercise that the lead team had begun.

The Final Piece

By early 1989 four or the five phases—familiarization, team network building, management systems and technical systems training, and strategic planning—were in place or being started at Androscoggin. The only piece missing was the introduction of statistical techniques as a means of more accurately defining manufacturing process weaknesses and of calibrating improvements. Some of the teams had already begun inquiring about or devising such techniques on their own to meet project requirements.

A member of the Corporate Quality Department, Robbin Guzowski, had previously worked at Androscoggin. She had headed the earlier effort to introduce Demings statistical quality control package there. She was one of those who had visited the Louisiana Mill. She had stayed in close touch with Belinda and had followed the progress of the team building and phase integration effort closely. Robbin was, therefore, the ideal person to provide the technical expertise necessary to the fifth phase.

Robbin, Belinda, and Tom met and mapped out a two-pronged strategy. The first prong involved training facilitators in statistical process control techniques so that they could offer this expertise when suitable projects arose. The second prong focused on finding at least one ongoing project in each department and showing how the techniques might be applied in order to improve results. The involved projects did not need to be team-generated. In fact, most of them were not.

By February of 1989, it was no longer necessary for Tom to spend time at the mill.

Topics for Discussion

1. What were the major differences between this start-up and the one at the Louisiana Mill?
2. How did the Androscoggin Mill manager's support differ from Bob Goins' and how did this difference affect the process?
3. What documents were generated to support the team effort and were they necessary? What others could have been generated and are they necessary?
4. How can safety programs actually be counterproductive?
5. How can a safety program and a systemic quality improvement effort support each other?
6. In what ways did the new management system developed at Androscoggin mimic the quality improvement process?

12 Culture as the Culprit: An Idealized Ending vs. Reality

After Reading This Chapter You Should Know

- What the qualifications of the head of a quality process should be.
- Why the Quality Improvement Department should be combined with Human Resources.
- Problems with the traditional Human Resources Department model and solutions.
- Why Human Resources trainers should be transferred to Quality Improvement and how their work should be organized.
- The importance of developing critical mass on the corporate as well as unit level.
- How to turn an organization quality improvement effort into a community-wide effort.
- Why the IP effort eventually failed.
- The ways in which dynamic conservativism can thwart a quality improvement effort.

The CEO Takes the Lead (Idealized Ending)

While the work was continuing at the Louisiana and Androscoggin Mills, IP as a whole was not progressing as rapidly as desired. Corporate emphasis remained on defining and putting into place the individual pieces. The corporate quality department staff spent a great deal of time on the road evaluating, training, and suggesting piecemeal improvements. As of yet, however, the department had shown no understanding of the need for a holistic approach, much less what such an approach involved. As a result, few manufacturing facilities and fewer headquarters support units had anything at all comprehensive in place.

At the mill level of the organization, employees assigned to head the quality improvement efforts had rarely received sufficient training. At the same time, they frequently had other, overriding responsibilities. Quality improvement teams were found in relatively few units. Where they were in place, a scattershot rather than a mill-wide, integrated approach existed. The teams were built around specific projects; or they were hourly teams headed by managers; or they were exclusively management-level teams. In most instances they were more like task forces then quality teams. And in no instance, other than at the Louisiana and Androscoggin Mills, had the necessary ground rules been adopted.

At this point in our idealized scenario, IP's CEO decides to tour the mills and see for himself how things are going with the quality improvement process. He has received glowing reports. In most instances, however, he expects that these reports are exaggerated. Two exceptions are the Louisiana and Androscoggin Mills. The tipoff here is that both mills have shown convincing bottom line improvements attributed at least in part to their systemic quality improvement efforts. The CEO saves these two mills until last. He spends two days at each. He sits in on team meetings at different levels and becomes familiar with the ground rules and their significance. He attends a lead team long-range planning session. He attends a facilitator meeting.

Several things quickly become obvious to the CEO during his visits. One is that while allowing his mill and corporate support unit managers to go their own ways initially might have been of value in terms of defining possible alternative approaches, it is time to institute one organization-wide approach so that everyone speaks the same language; so that common tools and techniques can be defined; so that a great deal of the competition can be eliminated between the mill managers and between support department heads; so that participants begin paying attention to and contributing to one another's efforts.

Another realization is that the Corporate Quality Improvement Department needs to be overhauled. When picking a new leader for this department, when redefining the job requirements, he realizes that engineering and management skills alone do not suffice. To produce something of value, future Vice Presidents of Quality need also an at least rudimentary understanding of the systems approach to quality improvement. The new Vice Presidents must be good with people. They must have individual and group process skills that enable them to deal effectively with employees at all levels, from the CEO right on down to the mill hands. But most of all, the new Vice Presidents need to be capable of developing the necessary integrated overview.

In terms of this position, the CEO also realizes that, rather than a training ground, the Vice President position should be viewed as a long-term commitment and a career track just as important as those in finance, production, and human resources. As a result, the next Vice President of Quality should stay at least until the desired model is in place organization-wide. Those who follow should have worked their way up in the quality field tract and should continue to implement and improve on what already exists, rather than starting over.

The CEO has learned that very few of IP's hourly employees and middle managers are truly committed to or have the understanding and skills necessary to contribute effectively to the process and that intensive training alone does not generate the required attitudinal changes, especially when it is completed in a classroom setting before employees have anything to relate it to. He realizes that the most effective training is on-the-job, with the trainees themselves helping define what should be included.

Finally, the CEO realize that some form of gain-sharing is necessary as a long-term incentive. Once employees have developed the necessary sense of commitment and have shown that they can help improve the bottom line on an ongoing basis, they expect and deserve to share in the new revenues generated by their additional efforts. Gain-sharing is another tangible way for Corporate to show its respect and thanks. It is also the best way to ensure employee commitment to long-term improvement.

Adjustments

On his return to Corporate headquarters, therefore, the CEO makes several announcements. The first is that Bob Goins will become the new head of the Corporate Quality Improvement Department. Mr. Goins has the desired engineering and management expertise. He has been with IP for 35 years and is well-respected. Finally, he is the senior employee most familiar with the systems approach to quality improvement that the CEO wants to put into place.

The CEO next announced that because of their contributions to bottom-line improvement, all employees of the Louisiana and Androscoggin Mills and of two other mills will be granted gain-sharing privileges. Payment will be made in the form of corporate stock instead of money in order to help remind employees that their ultimate objectives is to improve the performance of the entire corporation, rather than just of their individual mills, and to encourage an increasingly cooperative environment.

In line with the above, the CEO announces that bottom-line goals will be set for all other facilities. Upon reaching these goals, employees at those facilities will also be eligible for the gain-sharing stock program. He says that a formula will be created to allow facilities to share their gain-sharing rewards with corporate support units. The percentage contributed by each facility will be small, but as the number of facilities earning the right to gain-sharing grows, so will the stock bonus reaped by corporate staff units. This will encourage the latter to do everything possible to help improve productivity at the mill level.

Next, the CEO announced that the Quality Improvement Department will be combined with Human Resources. This will be a good move for both. It will give Quality more clout and access to more resources. The move will benefit Human Resources by allowing it to tie its activities more closely to the bottom line. By way of explanation, businesses can be defined in terms of five major functions—purchasing, production, finance/accounting, marketing/sales, and human resources. Each of the five has a well-defined focus. Purchasing, finance/accounting, and human resources control key inputs. Purchasing deals with raw materials and equipment; finance/accounting deals with money; human resources deals with manpower.

Traditionally, human resources has been responsible for employee recruitment, selection, evaluation, training and development, career management, compensation, labor relations, termination, and legal status. Traditionally also, human resources has received less respect than the other key functions. For example, it is normal for corporate presidents to have backgrounds in production, finance, accounting, sales, or marketing. It is unusual for an executive with a human resources background to rise to this level. At the same time, compensation levels and employee population levels are frequently increased or decreased based on financial, marketing, or production decisions with little or no attention being paid to the human resources viewpoint. Actually, in many companies, rather than being one of the five major functions, human resources has been relegated to the support category along with MIS and strategic planning. Rather than contributing to policy decisions, human resources personnel are expected to react to them.

The Problem with Human Resources

But if employees are our most important resource as more and more companies are beginning to claim, and if human resources is in charge of employees, how can this be? One of the main problems is that in our quantitatively-oriented corporate culture the human resources function lacks a bottom line. While purchasing can say, "See by how much we have reduced the cost of raw materials," or "Look at the savings this new piece of equipment will generate!"; while production can refer to quotas met; while finance and accounting can brag about how they improved earnings with their investments; human resources cannot document the fact that a new recruit will be worth $80,000 a year, or that a training program will increase employee productivity by 10 percent.

Human resources professionals are always having to prove themselves. Maybe this is why they work so hard. Maybe this is why they are so intent on interviewing as many potential recruits as possible, on training as many employees as possible, on offering as many people career counseling as possible, and on developing longer and more detailed evaluation forms. These efforts are evidence of human resources' desire to generate legitimate numbers of its own, though the sought-for recognition and respect rarely materialize. Human resources continues to come in last simply because its numbers, no matter how grand, no matter how potentially significant, do not translate directly into dollars, and, therefore, in the mind of quantitatively-oriented top executives, do not bear the same significance as those submitted by purchasing, production, finance, and marketing.

Human Resources as Key

When the nature of and the need for the five phase approach to quality improvement is understood, it becomes obvious that the human resources function is better equipped than any other to organize and facilitate the process. The success of at least three of the five phases described—familiarization, team building, and training—is dependent on getting a majority of the work force involved and committed. Also, the skills needed to facilitate and integrate a comprehensive quality improvement process are largely human resource skills—people and group organization, education, training, negotiation, and troubleshooting. The only critical skills that human resource professionals typically lack in this respect are systems skills.

Control of the quality improvement effort would provide human resources with a chance to directly affect the bottom line. If mounted correctly, such efforts can, as we have seen, help produce a substantial increase in corporate profits. Control of the effort would also give human resources more of a say in major policy decisions. Despite this win-win scenario, however, very few companies have coupled their quality improvement efforts with Human Resources in any fashion. More often, as we have said, the assignment has been given to a newly formed unit. This happens for three reasons.

First, upper-level management does not understand the true nature of holistic quality improvement well enough to realize that the key to success is employee commitment and to grasp the natural link to human resources. Second, the work done in most human resource departments is so specialized and so divorced from what is going on in other parts of the organization that department members are incapable of developing the breadth of perspective necessary to deal with quality improvement. Many actually see the demands of a full-scale quality improvement process as a threat to their well-defined, well-guarded domains. Third, at this point, human resource departments are too frequently amongst the most autocratic and hierarchical, with strict definitions of responsibility and authority. It is hard to help sell the participative approach necessary to an effective quality improvement effort to the rest of the company if you are not willing to practice it yourself.

Needed Changes

Clearly some changes are necessary if human resources wants to play a lead role in the quality improvement movement and reap the inherent rewards. One of these changes is that Human Resources staff must educate both themselves and upper-level management as to what a comprehensive quality improvement process involves so that the natural fit is understood.

A second change is that Human Resources training staff members must realize that their organization behavior, organization development, and psychology degrees do not give them all the skills necessary to contribute effectively to a comprehensive corporate quality improvement effort. The ability to work with and reorient individuals and groups is, indeed, important, especially during the familiarization and training phases. Relevant, organization-wide change comes most rapidly, however, if these skills are blended with the systems perspective, if improvements in products, production processes, management systems, and the work environment are focused on, rather than individual and group dynamics. When these systems are improved by a team-driven effort, the people problems frequently dissolve. Systems skills also give us the ability to effectively fit the pieces of the process together into a whole that is obviously more than the sum of its parts. Human resource trainers, therefore, must be willing either to develop the necessary additional systems perspective and skills, or to work closely with others who already possess them.

A final change is that Human Resource Department heads must be willing to introduce the team approach and more participative management to their own operation as part of the self-education process, to create a necessary model to point to, so that they cannot be accused of the "Do as I say, not as I do" contradiction.

According to the new arrangement, the CEO has decided that Mr. Goins, as Vice President of Quality, will work closely with the Vice President of Human Resources. He will also, however, report directly to the CEO. The Quality Department will offer five types of expertise—familiarization, troubleshooting, team building, group process skills training, and SQC/SPC skills training. Belinda Bothwick will be brought in head the team building section. Because of her familiarity with the approach being adopted, Robbin Guzowski will take the lead in the SQC/SPC training section.

Concerning group process skills, the Training Department of Human Resources will be folded into Quality Improvement. The head of training will report to Mr. Goins. The unit's charter will be altered drastically. Instead of defining corporate-wide training needs and delivering the training, department members will now be required to develop the flexibility necessary to respond rapidly to quality process-related requests.

By way of explanation, again, the CEO believes that an empowered work force, a work force spread all over the United States filling a wide range of responsibilities, through the vehicle of the quality team network, is much more capable of defining the majority of its own training needs on both the operational and the quality improvement side of the ledger than upper-level management and the corporate training department staff sitting in Memphis. The individual quality teams will define their units' needs. If these needs cannot be met on-site, they will be channeled upward to the Quality Improvement Department at Corporate. Staff here will either deliver the desired training themselves or help to find someone capable of doing so.

This does not sound like a radical change, but anyone familiar with the traditional corporate training model knows that it is. Normally, the head of training participates in the organization's long-range planning effort, or at least is briefed on the long-range objectives defined during that effort. This person sits down afterward either alone or with staff and defines the types of training that should occur during the next year to support the organization's objectives, as well as the resources required to provide this training. A schedule is set, classroom space allocated, time for training is negotiated with the heads of the involved departments, and the sessions begin.

Success has been measured in two ways, quantitatively and qualitatively. Quantitatively, the measurement has been based on the numbers of employees that have been run through courses during the year. Qualitatively, it has been based on student evaluations. The head of training has been totally responsible for the outcome.

This traditional approach to training is not the most effective in terms of organization improvement, especially in the new world environment. The problem is that it does not have the characteristics necessary to make it systemic. One shortcoming is that it is not truly participative in a systemic sense. Another is that it does not do enough to integrate the various organization functions. A third is that it is usually not ongoing and flexible in a systemic sense. And, fourth, a feedback mechanism that allows continuous learning as well as the improvement of the training model might exist, but is not comprehensive.

What we usually end up with, under these circumstances, rather than a effective training agenda, is an aggregate of pieces, of individual training sessions put together without enough attention being paid to the whole the organization wants to create. And, again, the responsibility for this failure is eventually borne by the head of training, who is trying to perform a very difficult task without seeking or being given access to the required support.

Let us go over the characteristics necessary to a systemic training effort one at a time. First is participativeness. Most trainers consider their approach to be participative. Employees are frequently involved in hands-on exercises. They break down into groups and practice a problem-solving technique, a creativity enhancing technique, a statistical measurement technique. They role-play and sharpen their conflict resolution skills. They actually work with a new piece of equipment. But the classes and exercises themselves are not where participation needs to begin to produce the best results. Instead, participation needs to begin at the real beginning. In order to be most effective, it needs to begin with the identification of organization, unit, and individual training needs.

Once the corporate strategic planning process is completed and long-term organization objectives are defined, top-level management has a pretty good idea what new skills should be learned, what old skills should be updated. But managers, in defining these training needs, are doing so with a top-down rather than a bottom-up perspective. Their reaction to, their interpretation of each unit's situation is usually quite different from that of the people who are actually in the trenches. It is the employees behind the desks or operating the machines who understand best which of their co-workers are having trouble meeting deadlines and why, which managers relate well to workers and which don't, why customers are complaining about the response they get when calling to place an order.

The Systems Approach to Training

In order to be participative in a systemic sense, a training paradigm must encourage the involvement of lower level workers as well as managers in the definition of training needs. It should also, when possible, allow employees to contribute to the design of the actual training modules. Finally, when realistic, it should encourage workers to actually help deliver the training themselves. From a systems perspective, therefore, we realize that participation must be incorporated into every part of the training paradigm, not just the delivery phase.

A second, major benefit of adopting such an approach is that it takes a great deal of the previously discussed pressure off the head of training. By making the paradigm truly participative, everyone shares the responsibility, everyone functions as a check on the validity of the training to be delivered. The head of training, in this instance, becomes a true facilitator, which, by definition, should have been that person's role from the start.

In terms of the second systems characteristic, the integration of training efforts across the entire organization, most companies again fall short. They do so because they leave this integration up to one already overworked person, our friend the head of training. When the systems perspective is adopted, a vehicle is created in the mills which allows representatives from all units to compare their training needs and resources, to see where these needs and resources overlap, and to help each other design the most effective delivery. This vehicle, of course, is the network of quality teams and the network of team facilitators.

With such an arrangement, the involved interaction also helps ensure that the organization training effort is ongoing and flexible. Teams are kept up-to-date concerning operational changes as a result of their participation in the lead team led long-range planning process. The ramifications of these changes in terms of training are understood more thoroughly in that each level—hourly, managerial, lead team—lends its own perspective. Also, when the teams continue to meet, changes in perspective can be brought to the table immediately and reacted to by everyone affected.

This new arrangement, of course, puts more pressure on the training department at all levels. According to the CEO's new design, the yearly training schedule becomes a thing of the past. The Training Department at IP must now become more of a resource, one created to meet the needs of the customer, as defined by the customer, on the customer's schedule. And in our idealized scenario IP's training department is no longer just three or four people struggling to stay up with the continually increasing demands of any progressive organization. Rather, it becomes a well-integrated, corporation-wide network of *employees* as well as professional trainers contributing to the involved decisions and responsibilities.

Finally, the feedback which allows continual learning as well as improvement of the training model is obviously a given when the organization-wide integrated network of quality teams is used.

Getting It Right

The CEO next decides that an organization-wide effort would be mounted to familiarize everyone with the chosen approach to quality improvement. The entire Human Resources Department staff will spend a week at either the Louisiana or Androscoggin Mill learning what the necessary pieces of the puzzle are and how they fit together. Afterward, the staff will help Bob Goins, Belinda, and Robbin develop a presentation. This presentation will initially be delivered to corporate executives during a two-day, "See, here is what we are talking about," seminar at one of the mills. Next, it will be delivered to mill managers and corporate support unit managers who will visit the two mills in shifts so as not to badly disrupt operations.

The last group to receive the presentation will be the heads of the quality improvement efforts at the different units, both operational and support. Finally, a video tape explaining what has been accomplished at these two mills and how it has been accomplished will be viewed by all middle managers and hourly employees. A question-and-answer session will follow each viewing.

Once the consolidation of the two departments had occurred, Belinda drew her staff of team building consultants from the Louisiana and

Androscoggin Mill facilitator networks. By this time, there were 30 to 40 trained facilitators in each. A good number could be spared without disrupting the process. Belinda sent two of her new staff to one facility in each division to help build the necessary team network or to reorganize an already existing one. The facilities picked were those considered furthest along in terms of understanding the newly adopted organization-wide approach and vehicle. These *pilot facilities* were then used as models and training grounds for the rest of that division. The main responsibility of Belinda's staff at these facilities was, once again, to help groom facilitators.

Belinda made a list of the consulting firms that had been used by different mills and the expertise that each offered. She then divided the list into two groups. The first included firms offering skills not possessed by members of the quality department/human resources staff. These skills included technical training and long-range planning. The second included firms offering skills that the combined quality department/human resources staff possessed. All eligible consulting firms were introduced to the systems model, told that they would have to fit their offering into it and that they would have to cooperate with each other.

A progress chart was started for each facility. When a facility wanted to use an outside consultant from the list, it had to clear its decision with the Corporate Quality Department. This requirement ensured that the five part progression was followed; that, for example, statistical techniques were not introduced before teams were in place and productive. It also helped keep facilities from wasting time and money on unnecessary training. Finally, it allowed the quality department to keep tabs on the effectiveness of consultants and to compare their costs.

Second Wind

The concept of critical mass is relevant on the corporate as well as the unit level. Efforts in individual units, no matter how successful, will eventually die if the process does not spread to enough additional facilities to become unstoppable. Quality improvement processes, until the very last piece is in place on both the corporate and unit levels, are either expanding and enriching themselves, or dying. There is no stable state. There is no halfway point at which what has been achieved can simply be maintained without further progress. This need for continual improvement and expansion is one of the reasons why a systemic perspective is critical.

When Mr. Goins left the Louisiana Mill to take over as head of corporate quality, one of his direct reports replaced him, thus ensuring process continuity. Another of his direct reports was made manager of a smaller sister mill, putting the key piece in place there, too. Within three years of the reorganization, a majority of IP's manufacturing facilities and corporate support units had mounted effective quality improvement efforts. Most manufacturing facilities were showing bottom-line improvements. The process ground rules first introduced at the Louisiana Mill had been accepted on the corporate level and were being absorbed into the organizational culture. For example, any unit now had access to any other unit for desired input or decisions. For example, anyone in IP affected by a unit's decision or project had to be asked for input and had to agree to changes before they were implemented on the operational as well as the quality improvement side.

If a unit was not receiving requested input from another, its head facilitator could appeal to IP's head facilitator, Mr. Goins. If he was also unsuccessful in dealing with the situation, Mr. Goins could carry the matter directly to IP's lead team, which included the CEO and all division heads for resolution. The lead team met periodically to review process progress, to receive unit presentations, and to lead the long-range planning phase, spelling out corporate objectives. The corporate lead team could also, of course, form a task force around any project it wished, using anyone it wanted to use, as long as that project had not already been started by a lower-level team.

Another challenge facing the revamped corporate Quality Improvement Department was to develop an organization-wide communication system. A centralized computer file was created into which all units fed projects that they were working on and had completed. Whenever hourly, management, or facility teams identified a new project, they were required to check the file, to see if any other unit had started or completed a similar one. If the answer was yes, they were obliged to contact that unit for input, thus potentially saving time while also establishing a new communication linkage. Representative facilitators from the units in each division met quarterly to discuss progress as well as process-related problems and innovations. Representative facilitators from the divisions met biannually to do the same.

Other positive things were happening. Requests from other companies started coming in for presentations on IP's approach to quality improvement. This occurred owing partially to improvements in product and service quality, owing partially to visits made by mill team members to familiarize themselves with the operations and requirements of both suppliers and customers, and owing partially to the fact that team members were using the ground rules to seek project and decision input from suppliers and customers who would be affected. When these presentations were made, several of the involved companies asked if IP could supply them with the expertise necessary to start their own process. The CEO thought this a good opportunity. The plan was that if the request came from a supplier/customer of the Louisiana Mill, for example, that mill would handle the presentation and send personnel to assist the process start-up. Corporate quality improvement would provide back-up, if asked.

At the same time, the mills began setting up employee speaker's bureaus. Mill personnel began networking with other local companies that had quality improvement efforts in place, and with companies of all sizes interested in starting one. Monthly brown-bag lunches were organized so that experiences and ideas could be shared. Videos were made describing the steps in the mill's process and were loaned to the community. Open houses were sponsored for local government officials, school administrators, and the heads of nonprofit organizations. The conversation at these open houses centered on defining how a quality improvement effort similar to the one at the mill might benefit them and how IP might assist.

As an increasing number of at least the larger IP facilities became involved in their communities, a model evolved. The local chambers of commerce, unions, and service clubs were encouraged to help organize and lead the ongoing familiarization phase. The educational sector took the lead in providing desired training. Courses were offered with mill and IP corporate staff in support. Finally, a long-range planning effort for the community or region as a whole was led by the academic sector in conjunction with corporate and nonprofit organizational planners. While the necessary critical mass had long since been achieved on the IP corporate level, the mills were now anxious to help achieve it on the community level. Other major corporations, impressed by the profits and prestige that IP was enjoying, began developing the same systemic perspective. As a result, they, too, became more effectively involved in community efforts to upgrade the overall quality of life.

What we are talking about in this last part of our idealized scenario is obviously quite possible. Community-wide efforts, indeed, have been mounted in many sectors of the United States. Most of them, however, have foundered for basically the same reasons that most organization quality improvement efforts have foundered. Leaders have not understood the magnitude of the desired undertaking and the need for a systemic perspective. Initial emphasis has been on training or on the introduction of statistical measurement techniques rather than on familiarization and team building as a means of rapidly getting people involved and of fostering the necessary commitment. The powers in the community have not been able to agree on one approach. These same people have decided that they are capable of putting the necessary pieces in place themselves rather than seeking professional assistance. Communities have become bogged down in defining what currently exists rather than concentrating on what ought to exist. They have become bogged down in developing tools to measure success before they have anything to measure.

The more successful community efforts that we are familiar with have characteristics similar to those of successful corporate efforts. These characteristics include:

1. A strong leader or a unified group of leaders.
2. A systemic rather than piecemeal perspective.
3. Professional assistance in defining and emplacing the appropriate approach and vehicle.
4. Up-front involvement of all key stakeholder groups in order to take advantage of their expertise and to gain their commitment.
5. An integrated network of project teams that facilitates the necessary free flow of communication.
6. Process ground rules agreed to by everyone.
7. Participative development of a plan for the community's future into which all other process pieces are eventually fit.

And that is it. That is the idealized version of what happened at IP. Next we shall talk about what *really* happened, which is a slightly different story.

Not Quite What We Had Hoped

Our idealized version of IP's future sounds rather straightforward and logical, perhaps nontraditional, but logical. It produced the desired results and, therefore, should have come to pass. As so frequently happens, however, logic did not prevail. Instead, it fell victim to what Donald Schon, in his book *Beyond the Stable State,* calls "dynamic conservativism." It fell victim to the efforts of those found in every organization who resist change because no matter how beneficial, it involves risk. IP went the way of most organizations. Despite continuing offers by both the Louisiana and Androscoggin mills to share what they had learned, the other mills, the Corporate Quality Improvement Department, the Corporate Human Resources Department, and Corporate headquarters in general remained uninterested. IP executives were extremely interested in the increased revenues being generated, but they seemed unable, or perhaps unwilling, to see the connection between these revenues and the systemic approach to quality improvement.

It eventually became obvious that Bob Goins' success at the Louisiana Mill was not going to provide a stepping-stone for him into the ranks of upper-level management. He was considered by too many superiors to be a radical and unorthodox. Bob eventually left the Louisiana Mill and was replaced by a corporate type whose first move was to make managers the leaders of the hourly teams. This new mill head was enforcing the philosophy, increasingly popular at corporate, that somebody had to be in charge. He also began assigning projects to teams and jumping the process, taking over team projects he found interesting. In a relatively short period of time, the Louisiana mill returned to its former position in the middle of the pack of mills in terms of profitability. At Androscoggin, the next manager, Larry Stowell, had worked with Doug at the Ticonderoga Mill. He had followed the efforts of OP&D and had watched the progress made at Louisiana. He was comfortable with the systems model. It, therefore, survived at Androscoggin. However, at the same time, it was not allowed to spread.

Dynamic conservativism is a cultural affliction. One of its major manifestations is an extremely autocratic management style. IP's CEO, John Georges, was, more than anything, an autocrat. He was a very traditional, extremely mechanistic boss. He made the important decisions. He reportedly had little interest in or respect for the input of those who worked under him. Their job was mainly to implement the changes he ordered. In terms of quality improvement, Mr. Georges wanted all the trappings, as well as the promised bottom-line improvements, but he obviously had no desire to see change occur in the way the company was run, especially at the top levels. Despite the encouraging speeches he made (at one point he warned that anyone who did not buy into the changes necessary for quality improvement would soon be looking for a new job) Mr. Georges' true sentiments were well known. As a result, most of those under him were unwilling, or, perhaps, were afraid to budge.

At the same time, partially as a result of the push toward improved quality, it became increasingly obvious that the organization's management philosophy was *the* main problem. Too much territorialism existed; too much hoarding of information as protection; to much back stabbing. Until this environment was dealt with, very little else would happen, leaving the people running the quality improvement process with the impossible task of making the necessary changes *without really making them.* They were faced with a true dilemma.

A second major problem was the organization of the Corporate Quality Improvement Department. At least initially, the Vice President of Quality position was seen, as we have said, primarily as a training grounds, a way to help rising stars gain the necessary overview of the corporation before moving on in about two years. This was the wrong emphasis and made progress extremely difficult, if not impossible. It was hard for anyone to learn all they needed to know from a systems perspective in two years, much less to begin an effective change process. The result, of course, was that, in order to make an impression, each new Vice President focused on one piece of a whole that cannot be divided successfully, a piece different, of course, from that chosen by predecessors, and to claim that this piece was the key one, that this piece would make the difference.

The chosen piece, of course, never did make the difference, but this propensity of the new Vice Presidents kept the process in a stage of constant flux. Quality Department staff and those running actual projects in the field were required periodically to radically shift their perspective. This was extremely discouraging and disconcerting, to say the least. Headquarters demanded that they pay attention, but then headquarters kept changing its mind. Eventually, they just began to play along, going through the motions, mounting the required training programs, but doing as little as possible beyond that because they knew that in a year or so it would all change again.

An off-shoot of this prevailing flavor of the month attitude was that no concept of what, exactly, was required for a successful QIP materialized. Crosby had given the company a lot of ideas, buzzwords, tools, and techniques; but he had given it no integrating concept. As a result, the process ended up being treated basically like a traditional incremental training exercise, designed by top-level management and delivered top-down. What needed to be learned was gained from the classroom or manuals in pieces, and the pieces never made up the necessary, integrated whole.

A third major problem was the competitive environment that prevailed at IP. Managers at all levels competed and were forced to compete against each other. At the mill level production figures were constantly compared. The object was to beat the other mills. This was how managers gained status. At the executive level the sole objective was to please Mr. Georges. One's job depended on it, so that the various executives and unit heads were focused totally on their own domains, their own problems, and their own strategies. They did not hesitate to reshape the quality effort in their own image, trying to find a way to do things better in order to stand out from the crowd. The Crosby model had given them very little to work with when it came time to actual implementation, so that separating one's self from the pack was easy. The question was, what direction to go in? Bob Goins was brave, took a chance, and lucked out, at least in the short term, but most mill managers had absolutely no clue.

Unfortunately, this is the kind of challenge that a majority of employees in U.S. firms are faced with. To compound the problem, there are still relatively few like Bob Goins, Ralph Stayers, and W. L. Gore around. The prevailing culture is still very much so top-down and competitive with people at the top clinging to their hard-earned power. Occasionally, someone sees the light and is willing to take the involved chance, but most such people are either new and looking for a way to attract attention, or are extremely secure in their positions and looking for some excitement in their lives. Those who are new tend to get fired or beaten back into line. Those who are secure and looking for excitement tend to get isolated.

This is why the Crosby model and others like it remain so popular. They are top-down, management-driven, and tightly controlled. Nothing really moves unless management says that it can. This is why the systems model has not been adopted by more like Goins, Stayers, and Gore. The refrain, "Somebody has to be in charge," still rings through the halls of the executive suite, and is emphatically reinforced by evaluation and reward systems, by organization structure, by MBA programs, by the way training is defined and delivered. These are the things that shape the culture of organizations. Until they change, the culture is not going to change. The only person capable of precipitating the necessary change is the person at the very top of the pyramid. Bob Goins was at the top of his pyramid and fostered the desired change at the Louisiana Mill. Unfortunately, his pyramid was overshadowed by a much larger one, and the person at the top of that pyramid was not interested; so that no matter what happened below, the necessary organization-wide awakening was not going to occur.

One last demonstration of the fact that culture was the culprit at IP is that after the Louisiana Mill had become the most profitable mill in the system, it was suggested that the employees there be rewarded financially for their contribution, that some form of profit-sharing or gain-sharing be instituted. Corporate said, "No." Its rationalization was that if Corporate did it for the Louisiana Mill, employees at the other mills would begin expecting the same treatment. This is a classic example of what Russell Ackoff, one of the fathers of systems thinking, calls reactive thinking, of dealing with change, with the possibility of change, solely from the perspective of a traditional paradigm. Underlying this decision, of course, was the need to keep the employees in their place, the fear that if any concessions were made the employees would immediately begin demanding more. This is the same fear that keeps management from allowing employees on teams to identify their own projects. The fear is two-fold. First is the suspicion, which in our experience has proven untrue, that hourly employees especially will focus solely on selfish interests and will show no concern for the welfare of the company as a whole. Second is the concern that if they are empowered, if they are encouraged with monetary incentives and do create positive change, the employees will become uncontrollable and will no longer see the need for management in the traditional sense.

The second fear, of course, is extremely valid, because management in its traditional role is, indeed, obsolete and increasingly ineffective. The problem, once again, is culture. We tend to think in win-lose, reactive terms which translates to, "Either we have traditional managers, or we don't have any managers at all," rather than in nontraditional, systemic terms which translates to, "It is time for the role of management to evolve."

One last comment: It is interesting to note that while IP's executives were saying no to the mill employees they themselves were receiving bonuses based on the company's bottom line. The executives were benefiting from the mill employees' improved productivity while those directly responsible for it were not allowed to. Finally, it is interesting to note that despite this setback, the work force at the Louisiana Mill continued to set records, or they did so at least until the new, more traditional manager replaced Bob Goins.

Corporate culture, dynamic conservativism, reactive thinking, the inability to accept change; these are the enemies of quality improvement, not only at IP, but at a majority of organizations. This is what the real gurus, the Demings, Ackoffs, Jurans, and Peterses, are talking about. This is what most people have trouble hearing. The problem, of course, as we have said, is most severe at the top. This is why we shall focus on top-level management in the next chapter.

Topics for Discussion

1. Why are most companies reticent to introduce a reward system that ties everyone's salary to the bottom line?
2. What skills are important to success as the head of a quality improvement effort?

3. Why should the quality improvement department and human resources be closely linked?
4. How does your organization's training model compare with the systems model?
5. List the things that the IP CEO did which benefited the quality improvement process in the idealized version.
6. What actions can an organization take to encourage a community-wide quality improvement process?
7. Give examples of dynamic conservativism in your organization.
8. How does the failure of the IP effort prove that the person at the top of the organization is the most important player?

13 Top Level Management as the Key

After Reading This Chapter You Should Know

- Why most organizations are not qualified to organize a successful quality improvement process.
- Why some CEOs pay themselves too much and the consequences.
- The value of tying everyone's reward to the bottom line in some fashion.
- Why union leaders hesitate to support quality improvement efforts.
- Why traditional union tactics no longer bring the desired results.
- Changes unions should make to become effective leaders in the quality improvement movement.
- How unions can keep a quality process headed in the right direction.

Good Guy, Bad Guy

The major in-house roadblock to mounting a comprehensive quality improvement process has been defined by many as middle managers. Middle managers feel threatened, for the reasons previously explained, and resist. Middle managers can, indeed, be an obstacle to an effective quality process, but this can be overcome if employment security is ensured and if their new role as facilitators and change agents is properly presented and reinforced. The most serious obstacle is not middle management; it is, in fact, senior management, the people upstairs who kicked it all off and who appear periodically to make well-crafted speeches stressing the fact that improved quality requires cultural change and must become a way of life.

The trick, of course, is to watch their feet as well as their mouths. Employees do just that and learn all too frequently that upper-level managers are indeed for improved quality and the necessary changes, but only so long as they themselves are not affected, only so long as alterations in their own styles of management are not necessary. They have neither the inclination nor the time. They are currently involved in too many crises upon which the fortunes of the company depend to worry about changing the way that they do things.

When upper-level managers feel this way, failure is usually assured. The effort might drag on for years and produce some noteworthy results, but ultimately it will not meet expectations. If senior executives do not set the example and play by the rules, no one else will. If the top people decide that they are allowed to modify the rules in order to deal with the pressures of leadership, others will quickly follow suit.

The executive corps of any corporation has both long- and short-term objectives when it invests in quality improvement. The long-term objective is to steadily improve the corporation's bottom line through better planning, relevant training, the introduction of appropriate statistical measurement tools, and through better use of employee expertise. The short-term objective is to enhance the company's image by publicly hyping its new dedication to improved quality. Unfortunately, too many of those that start out with both long- and short-term objectives end up focusing on the latter. They become more interested in creating the image of improved quality than in actually improving it. This happens because the executives eventually realize that they do not possess a comprehensive enough perspective to engineer what is necessary to positively impact the bottom line through quality improvement.

One reason that they do not possess the necessary perspective is that a majority of them are quantitatively oriented. They are finance people, accountants, engineers, and lawyers. They are most comfortable when they can reduce things to numbers: return on investment, cash flow, staffing levels, working capital, return on net assets, and so on. This helps explain why Deming's quantitative tools are so quickly adopted by so many as the all-inclusive answer to quality improvement, while the rest of his message is ignored. It explains why we recently saw a poster on the walls of a corporate headquarters announcing that productivity is simply a matter of the right numbers and was told that the CEO himself had boiled the entire quality improvement effort down to this one line. Finally, it explains why so many executives think it necessary to invest sizable amounts of effort and money up-front to develop quantitative tools for measuring process success when experience has proven such tools to be largely unreliable and unnecessary.

Sharing the Blame

Executives who lack the necessary perspective lack it, at least partially, because they have not been properly trained. Executive training comes from four sources—colleges and universities, in-house programs, consultants, and on-the-job coaching. Despite the growing chorus of pleas from both the private and public sectors, most of academia continues to focus on enhancing quantitative and other technical skills. Those skills necessary to developing employee potential, to integrating employee efforts, and to achieving an effective overview of the total operation and how it fits into the economic, political, and social environment of which the organization is a part are too frequently ignored or skimmed over.

In-house training and consulting packages can be of value, but are too limited in duration and scope to ensure the necessary changes in attitude and culture. On-the-job coaching by bosses should provide most of what is necessary. A majority of bosses, however, have not been adequately trained themselves, or are too busy, or see their reports as competitors and shy away from sharing expertise.

A second reason that quantitatively-oriented executives frequently do not possess the necessary perspective has to do with the pressure that the financial world places on corporations. The head of the organization is ultimately responsible for divining the pending impact of environmental forces and for preparing the organization to profit from that impact. In order to do so, the CEO needs to keep an eye on both the environment and the organization. Ideally, the CEO can concentrate on the organization as a whole, while different parts of that organization monitor and make valuable recommendations concerning environmental forces.

The snag is that the U.S. industrial sector is supported mainly by stock investments handled by brokerage firms whose customers are interested primarily in short-term gain. CEOs are too frequently forced to pay a disproportionate amount of attention to this one segment of the external environment. The situation has been made worse by the threat of corporate raiders. As a result, boards of directors lean toward candidates with financial expertise and insist that their CEOs pay close attention to what is happening in financial centers. Such expectations add to the quantitatively-oriented executive's clout and help excuse his or her lack of attention to in-house, nonquantifiable issues.

Taking this into account, we cannot fault top-level executives entirely for their quantitative orientation and for their failure to understand and play the role necessary to a successful QIP. Wall Street also needs work. Brokers are too often engrossed in their own reality, one strictly of numbers, one devoid of the faces and emotions necessary to improve quality. Too many aggressive financial predators are popping up. Too many intelligent and highly creative young men and women are sitting in their chrome and glass towers trying to figure out ways to beat the system in order to do better than their peers. Competition/conflict is alive and well in the world of finance, maybe more so than anywhere else. Life for too many brokers and financiers has become an "in-group" battle divorced from the reality that the decisions made and the victories won on Wall Street create elsewhere.

Finally, the big-time money manipulators are hurting the quality improvement badly. Leverage and other types of buyouts can be effective quality-enhancing tools in poorly managed companies frustrated by bureaucracies or frozen by tradition. But the new owners must be willing to make the necessary changes and to follow through. Many of today's takeover artists have shown no such willingness. They just want to drain the assets of the company, then dump it and move on, showing no concern for the effects of their greed on employees. They are worse than the original robber barons. The originals, no matter how inhumane their tactics, at least created something of value to society—railroad systems, the steel industry, and the oil and banking industry—with their efforts. Our modern day version, frequently, shows no such desire. With them it is all take and no give.

Management's Achilles Heel

Another important reason that top corporate executives are the main roadblock to successful quality improvement efforts is their inability to function as team players. They might talk about their team and go through the motions of seeking input from reports, but most decisions are ultimately based on their own intuition or judgement. This characteristic is well-documented by Robert Lefton and V.R. Buzzotta in their article "Teams and Teamwork: A Study of Executive-Level Team." The authors examined 26 top-level corporate teams, 20 of which headed Fortune 500 companies. The researchers found that little effort at real communication existed, little listening, a lot of grandstanding, a lot of turf battles, little participative decision making, and little interest in the problems of implementation.

Another indication that many top-level executives are not team players is that they pay themselves too much and refuse to tie their compensation to the bottom line. Measurements of the reasonableness of executive compensation include comparison of the involved yearly increase with return on the average shareholder equity, comparison with corporate profits, and comparison with the increases that other executives in the same and other industries receive. It is difficult to find a model that compares what top-level executives make with what other employees in the same organization make.

In terms of tying compensation to the bottom line, we have all read about companies that are not doing well where no annual compensation increase has been offered, or where everyone has been forced to take a cut. Everyone, that is, except the top-level executives, who have found reason to award themselves bonuses or salary increases. Such moves do not foster loyalty to the organization, to the team, or to a quality improvement effort. As a student who works for a bank recently wrote:

> "During the last two years, our company's main industry has been extremely competitive and margins have been sacrificed for market share. Owing to the fact that, during this time, our share of the market has not grown appreciably, upper-level management has been forced to initiate other steps to improve the bottom line. One such step has been the elimination of merit increases. During this same period our upper-level managers have received record high bonuses, a strategy which has not delighted too many employees. It would seem reasonable to cut back on these bonuses or to eliminate them entirely, as the involved individuals already receive 10 to 20 times the salary of the average employee. A compensation system that ties every employee's share to the bottom line would improve employees' perception of the fairness of the system."

Actually, the 10:1 or 20:1 ratio cited does not even come close when referring to the compensation received by a growing number of CEOs. The authors of an article in the May 1, 1989 edition of *Business Week* entitled "Is the Boss Getting Paid Too Much?" surveyed 254 companies and found the average CEO's compensation in these companies to be 93 times that of the average hourly worker. Since then, that ratio has continued to increase, until today it has been estimated to be at 125:1. The rationale supporting such a high level of compensation is threefold:

1. The CEO works harder and makes more difficult and more critical decisions than everyone else.
2. This high level of compensation is needed to attract and to keep superior talent at the top.
3. In order to protect the company's image, the CEO's pay level must remain competitive with or exceed that of other companies in the same industry.

In terms of the first argument, if the CEO works harder than anyone else, it is because that person is not a team player and feels the need to make all the decisions. A good, new world corporate leader does not. A good leader gets the right people in the right positions, encourages them to make the decisions, supports them in their efforts, and helps integrate. The person that we are talking about, then, is a boss rather than a leader and has only himself or herself to blame for the fact that the company seems unable to progress beyond crisis management.

In terms of the second argument, there is little value in having the best and the brightest at the helm making decisions if those below, those responsible for implementation, feel disgruntled and are saying, "Let the person who is making all the money solve the problems."

In terms of the third argument, the relationship between the extremely high compensation level of some CEOs and the company's image is not necessarily a positive one. The current rash of articles in major magazines questioning the value of such packages provides evidence of the doubt in the public mind. The people who care the most, of course, are the company's employees. They generally see the comparable worth rationalization exactly for what it is and wonder how the CEO finds the nerve to ask them to be less selfish.

More Lessons from Abroad

Top-level executives who understand the need to mount successful QIPs but who continue to pay themselves too much and refuse to tie their compensation in any way to the bottom line still have important lessons to learn from the same Europeans and Japanese who helped convince us of the need for improved quality in the first place.

According to Jan Wessels, the deputy consul general of the Netherlands in New York City, the ratio of CEO pay to that of the average factory or office worker with ten years of experience in industrialized Western European nations is between 8:1 and 9:1 before taxes. What he is saying is that if the average employee is making $20,000 per year, the CEO is making between $160,000 and $175,000, instead of close to $2,000,000 as in the United States. In some of these countries, the ratio is enforced by law. The CEO is allowed to make as much as he or she wants, but, at the same time, that sum can be no more than 10 or 20 times what the lowest paid employee in the company makes. This arrangement forces a team approach. Corporate leaders can no longer manipulate organization systems solely for their own benefit. It is now necessary for them to encourage the development of employee potential on all levels.

According to Japanese executives contacted by David Sulz, a staff member with the Wharton School Japan Program at the University of Pennsylvania, this ratio lies between 12:1 and 17:1 when comparing the salary of CEOs with that of new hourly employees in Japan. Also, according to an article by Katsusada Hirose entitled "Corporate Thinking in Japan and the U.S.," when profits fall, Japanese executives and board members, as leaders, are more willing to ask the workers' forgiveness and to take the initial compensation cuts themselves. For example, according to *The Japan Economic Journal,* December 20, 1986, because of losses resulting from poor overseas ventures, Fujitec Corporation slashed executive salaries by 10 to 50 percent. According to the *Japanese Times,* December 31, 1987, Nikko Securities canceled a planned yearly pay raise for executives because a financial deal had gone sour and the company wanted those responsible to search their souls. According to the *Nikkei News Bulletin,* April 4, 1986, Mitsubishi Electric, Fujitsu, Toshiba, and Sansui Electric all cut board members' salaries. Exports were lagging owing to the growing strength of the yen, and profits were down.

Presently, the Japanese are experiencing difficulties. They are going through what we call their robber baron phase. They have collapsed the industrial development cycle that took us approximately 100 years to complete into roughly 40 years. Their traditional focus on the good of society has temporarily been distorted by too much growth too fast. They have lost their perspective in terms of values; or perhaps some of their traditional values are now necessarily being challenged, but their management philosophy remains sound and one of the most effective around. It is fairly well guaranteed that the Japanese will pass through this phase fairly rapidly and will move on.

The Right Way

Another characteristic of too many top-level executives is that they do not believe in sharing the rewards of improved quality with the employees who have contributed to the involved improvements. Their argument is that if employees are paid a competitive wage and are given decent benefits, it is their responsibility to do the best that they can for the company and to expect nothing extra in return. The problem with distributing a part of additional profits is that employees will begin to expect such rewards whether the company does well or not. Also, the unions will immediately attempt to turn such gift giving into a precedent and will begin negotiating it as a permanent part of the contract.

In contrast, firms seriously interested in improving quality eventually tie everyone's compensation to the bottom line. Everyone who contributes gains a share of the increased profits generated. This share can be in the form of pay raises, bonuses, profit-sharing, gain-sharing, or of employee stock ownership plans. No matter what form it takes, however, it is an essential ingredient.

One reason that some U.S. companies have soured on the sharing of profits is that they have introduced their plan at the beginning of a quality process as a means of stimulating interest, rather than waiting and delivering it as a reward for impressive bottom-line improvements. The results of this strategy have been poor. Successful quality improvement efforts generally follow a pattern. The first phase answers the questions:

"Why quality improvement?"
"What is the purpose of the team network?"
"What kind of training can we expect?"
"What is the role of planning in all this?"

It generates curiosity, skepticism, and some enthusiasm in the work force. During the team building and training phases that come next, workers and supervisors get excited about the fact that they have been given the authority to recommend, design, and implement improvements. This excitement is reward enough initially and is supplemented by the more cooperative atmosphere engendered.

In the following months, the payoff becomes the steady improvement seen in overall productivity and in the bottom line. Everyone enjoys being on a winning team. Eventually, however, quality process participants on all levels begin thinking and saying, "Hey, we have helped improve the fortunes of this company. We should share in the financial rewards." It is this point, then, that the customized plan should be introduced. Such timing will further heighten employee morale and give a boost to the process. If structured right, it will also provide an ongoing incentive. For example, we have never heard of flagging interest in improved quality at the previously-mentioned Lincoln Electric, the much talked about Cleveland-based manufacturer of arc welding equipment and induction motors, where, according to the article "This Is the Answer" in the July 5, 1982 issue of *Forbes,* employment is guaranteed for those with 2 years or more of service and where up to 50 percent of yearly compensation frequently arrives in the form of a bonus based on performance.

In sum, we have all heard about the need for cultural change if we are to achieve improved quality. Only now, however, are we beginning to realize the full implication of this pronouncement. Yes, the workplace has to grow more participative and less authoritarian, but we have not yet gone as far as we need to. The only people with the power to lead the way are our top-level executives. But in order to do so successfully, many must first work on their own mind-set, their own image. They must achieve a more balanced quantitative/qualitative perspective. They must become real team members. They must be willing to tie their fortunes to the bottom line along with those of the employees. Only then, bolstered by the support that the team atmosphere provides, will corporate leaders realize their true potential and power. Only then will they become capable of fostering the desired cultural change, both within the workplace and outside. Only then will the quality improvement that they advocate become reality.

Union Leaders as Well

Unions have traditionally led the way toward an improved quality of working life. Historically, they have played a major role in securing the 40-hour work week, child labor laws, minimum wage, protection from injury and harassment, and fringe benefits such as health care, vacations, sick leave, and pensions. In terms of the modern-day quality improvement movement, however, as John Hoerr wrote in his article, "The Payoff from Teamwork," it is now management (more so than union leaders) that is beginning to push employees in both union and nonunion plants to accept more involvement.

Several reasons exist for the hesitancy and caution of union leaders. One is that although they have as much to gain in the long run, they are more at risk than CEOs. Specifically, while CEOs can fake commitment to a quality improvement process, union leaders cannot. CEOs can be extremely gung ho up front, spend millions on consultants, and mandate the involvement of supervisors and hourly employees. At the same time, however, as we have said, they can refuse to make the changes we have talked about in their own styles that are critical to success. CEOs can focus on the image of improved quality while avoiding the content and get away with it, at least in the short run. The work force has little leverage, even when the CEO's lack of commitment becomes obvious. At the same time, board members and stockholders are too far removed to evaluate actions properly.

Union presidents, on the other hand, are directly responsible to and dependent on those most deeply affected by the work-life changes being advocated. They are dependent on the votes of union members to stay in office. They must be sure that quality improvement is not just a management plot to reduce the union's power, or another efficiency study in sheep's clothing aimed at getting more out of a pared-down work force for less.

Another reason for the relatively slow uptake of unions concerns the difference between corporate and union politics. CEOs can force superintendents, supervisors, and foremen at least to give the process a chance. Promotions and, in extreme cases, continued employment can be used as incentives. National union leaders, however, cannot apply such pressures to local presidents. The latter, as we have said, are elected from below rather than appointed from above.

Finally, local union presidents frequently lack the broad perspective of CEOs. Their attention is focused on office or shop-floor issues. The fact that the company is losing world-market share to cultures that have developed a more supportive labor-management relationship—that the traditional adversarial stance is outdated—can be hard for them to relate to.

Just Cause

National union leaders can apply some pressure to get local leaders to accept and contribute to quality improvement efforts, but they must remember that the presidents and members of their local unions know better than anyone what is really going on behind the quality improvement banners at plants and offices. They are naturally suspicious. They have good reason to be. In the United States, corporate management's history of duplicity in its dealing with labor has been well-documented. The current trend toward paying closer attention to workers' ideas and needs as a means of enhancing both the quantity and quality of output is nothing new. During the last 30 years, movements in this direction have cropped up periodically, especially during down cycles in the economy.

But while job security has been the number one prerequisite to success in similar European and Japanese efforts, the U.S. sector, despite its supposed understanding of this critical relationship, is currently undergoing yet another wave of corporate staff cuts. One reason for U.S. management's inability or unwillingness to make job security a number one priority given in an article by Janice Castro entitled "Where Did the Gung-Ho Go?" was, curiously enough, the pressure of increased global competition. Others reasons included the desire to avoid unfriendly takeovers; the current, unprecedented level of corporate debt; and pressure to show constant improvement in the bottom line.

Union members witness the current cuts in benefits being negotiated, the increasing use of temporary workers with no benefits at all, the increasing use of robots, the comparatively small amount spent in this country on training to enhance employee skills and opportunities, and they have trouble believing that the purpose of the quality improvement process is to better *their* situation as well as the profitability of the company. The fact remains, however, that:

1. Increasing numbers of our unprofitable companies are being bought and made profitable by foreigners who send the profits home.
2. If this situation does not change, workers and unions as well as management will lose.
3. The QIP vehicle is the best that we have for defining and implementing the changes necessary.

A New Face

Unions have a chance to make the difference. They have the power to help U.S. companies move more rapidly in the right direction. First, however, they need to do some rethinking and reshaping of their own role. They need a new face. The old one is outdated. It no longer attracts the way it once did. Unions need a new strategy for achieving their ultimate ends. Strikes do not work anymore. Management has developed the legal and economic clout necessary to beat or wait strikes out. As a result, according to the Bureau of Labor Statistics, the number of labor stoppages involving 1000 or more employees has dropped from an average of 331 per year in 1950s, to 299 per year in the 1960s, to 269 per year in the 1970s, to 64 in 1984, to 54 in 1985.[1]

The more recent tactics of smearing corporations by spreading stories about them and of boycotting other organizations—banks, suppliers, customers, and so on—that do business with the target are not working either. Such efforts divorce themselves largely from the union's traditional source of power—the workers. They move the action onto the corporate playing field. Union lawyers and public relations experts trade blows with corporate lawyers and public relations experts while those who started the whole thing, those whom the battle is supposed to be about, spend most of their time watching from the sidelines.

What we have here is a race of sorts. If corporate management continues to lead the way toward comprehensive quality improvement, unions might end up even worse off in terms of declining membership. If unions pick up the pace, however, they can share leadership in one of the most important social and economic movements of modern times. A growing number of progressive union leaders are realizing this and are demonstrating bursts of speed. Lynn Williams of the United Steel Workers (USW) became a strong advocate of a cooperative labor-management approach to quality improvement.[2] Owen Bieber of the United Auto Workers (UAW) was been quoted as saying, "Unions must be willing to establish joint programs with management on shop floors to staunch the flow of jobs overseas."[3] When asked in an interview with *Pulp and Paper* if unions should not take the lead in improving labor relations and working for better quality, Wayne Glenn, president of the United Paper Workers International Union (UPIU), replied:

> "We have taken the initiative. We've tried to teach our staff how quality improvement programs can be used for the good of all. We want to keep workers from thinking that the company is trying to undermine the union with these programs."[4]

At the same time, strong resistance can be found among the followers of all these leaders. Local 1010 USW president Mike Mezo said, "We don't think there's any benefit to cooperation. No way will we ever take part."[5] A New Directions faction broke away from the UAW majority to fight the cooperative movement, saying, for one thing, that it has done little to ensure job security.

In 1988, a book entitled *Choosing Sides: Unions and the Team Concept* appeared. The authors' advice was to "Just say no!" They equated the team concept with management by stress. They called it a subtle effort by those in charge to ease workers into an even more demanding form of Frederick Taylor's efficiency-seeking scientific management.

> "Management by stress (the team concept) uses stress of all kinds—physical, social, and psychological—to regulate and boost production. It combines a systemic speedup, just-in-time parts delivery, a strict control over how jobs are to be done to create a production system which has no leeway for workers, and very little breathing room."[6]

The authors saw collective bargaining agreements that included the team concept as heading workers toward the following:

1. Interchangeability, requiring or inducing workers to learn several jobs.
2. Drastic reduction of job classifications giving management increased control over the assignment of work.
3. Less meaning for seniority.
4. Detailed definition of every step of every job, again increasing management's control.
5. Worker participation in increasing their own work load.
6. More worker responsibility without more authority.
7. An attempt to get away from the "I just come to work, do my job, and mind my own business" attitude by showing them how they fit into the whole operation.

Much of what the book says is true, but the authors, unfortunately, have interpreted the quality improvement movement according to their own extreme bias, leaving out the parts proven in most cases to enhance rather than degrade the employee's quality of working life. They have also said that while responsibility is increased by a quality improvement process, decision-making authority is decreased. If the systems approach to quality improvement is used, this statement is absolutely untrue. The book, however, and its popularity, indicate the level of distrust that has to be overcome in some quarters.

Actually, many of the problems facing today's union leaders are similar to the ones faced by CEOs. These include an inflexible, hierarchical union management system; too much bureaucracy; fear of change on all levels. To improve the effectiveness of their operation, union leaders should take many of the steps recommended for CEOs. These include:

1. Begin a campaign to familiarize membership with the concept of quality improvement as it applies to union management and operations. Start pushing more problem-solving and decision-making down to lower levels.
2. Develop a team approach to the management of unions (circular organization) complete with ground rules. The teams in the network could be built by region or by company.
3. Mount, with assistance from academia, a training effort to provide union members with the skills necessary to function effectively in a team-driven atmosphere.
4. Turn national and local leaders into facilitators, teachers, and resource generators, rather than decision-makers and problem-solvers.

Another sacred cow that needs to be sacrificed in most instances is work rules and job restrictions. According to Peter Drucker in an article entitled "Workers' Hands Bound by Tradition," it is exactly those industries most tightly bound by work rules and job restrictions—steel, automobiles, consumer electronics, rubber, and so on—that have done the poorest against foreign competitors. Work rules and job restrictions were developed to protect jobs. At the same time, however, they have made jobs even more repetitious and boring, have allowed employees to realize only a limited portion of their potential, have produced a team where members cannot assist or fill in for missing teammates, and, in the end, have proven self-defeating in that the resultant loss of productivity has led to layoffs and shutdowns.

More Powerful Than Before

The strategy of taking the lead in the quality improvement movement can be a no-lose opportunity for unions. Their traditional concerns—job, security, reasonable pay, safety, improved work environment, and benefits—match those of a systemic quality improvement approach. At the same time, this strategy will improve their status in relation to that of corporate management. Unions are the only sector capable of forcing CEOs to take process-related responsibilities seriously. They have the hammer necessary to make corporate executives play by the rules instead of creating their own or changing the rules whenever convenient. That hammer is their access to employees who can document shortcomings in combination with access, through the media, to a public increasingly interested in how our corporations are being managed or mismanaged.

Blowing the whistle on a CEO who is faking a quality improvement effort can affect both the company's bottom line and the CEO's career. It can provide a missing but much needed public service. At the same time, it can help clarify the union's role as guardian of the customer, the stockholder, and the nonunion as well as the union employee.

The sacrifices that union leaders will have to make to prove their sincerity in this new role—getting membership more involved in the management of the national through a network of teams, softening or doing away with work rules and job restrictions—should not be a giveaway. Something should be demanded in return. Possibilities include a more equitable salary/bonus/ benefit system that ties everyone's take to the bottom line, and a role for unions in the long-term corporate strategic planning process.

The division of profits is a hot issue. The public is becoming increasingly incensed over the salaries, bonuses, and benefits that CEOs are awarding themselves. Unions could throw their strength into an effort to force CEO rewards back to a sensible level and to tie not only the CEO's but everyone's take in some way to the long-term bottom line. Profit sharing, gain sharing, employee stock option plans (ESOPs), and employee ownership are some of the vehicles that have been developed. Implementation of these vehicles, however, has not always been smooth. Top-level management has balked or misrepresented the facts. Lower-level employees have had difficulty understanding the benefits of such a change, or have interpreted it as another management trick. Unions, taking advantage of their rapport with workers, could help define and implement the most appropriate vehicle for each situation.

Beginning at Home

Meanwhile, on the local level, in their new role as facilitators, unions leaders can help guard against workers improving themselves out of a job. As a result of increased productivity and the improvements in supplier and customer relations, a comprehensive QIP usually ends up causing more people to be hired than fired. Job definitions, however, change. Some jobs disappear. Others are combined. Short-sighted number crunchers cannot be allowed to use these changes as an excuse for layoffs and the subsequent, unreasonable increases in responsibility for those remaining.

Unions must also help protect team participants from the wrath of threatened middle managers. Upper-level managers are sometimes hesitant to step in and stop such harassment. They do not want to further alienate a valuable and loyal middle manager. They feel sorry for Joe or Mary and shrink from further aggravating his or her frustration. They might even secretly agree, allowing the manager to say and do what they are unable to. Whatever the reason, union representatives can function as watchdogs, helping ensure that upper-level management reacts properly to such process-threatening behavior.

Unions can help see to it that quality improvement process participants receive proper training. Team members usually do a good job of identifying their own training needs. However, many corporate training departments, as we have said, have their own agendas and are not organized to meet needs defined outside those agendas. Unions can either play the advocate for workers and encourage corporate training departments to adapt, or, in conjunction with academia, they can help employees establish their own training programs. Finally, union leaders can continue to educate themselves and can help educate industrial leaders about the need for a systemic rather than a fragmented or totally lop-sided approach to quality improvement.

In summation, the quality improvement movement that some union leaders and members still see as a threat can also be regarded as a tremendous opportunity. What it requires is a new perspective and the willingness to shift priorities. The involved changes will not be easy. If unions can make them, however, they will ensure a key future role for themselves in the U.S. manufacturing and service sectors. They will also help hasten a shift from our worn-out, confrontational labor-management posture toward a more comfortable and productive win-win one.

Topics for Discussion

1. How sincere is your organization's CEO concerning quality improvement?
2. How effective is your organization's executive corps in leading your quality improvement effort? How can it improve on its performance?
3. Does your organization's reward system encourage participation in the quality improvement process? How can that system be improved?
4. What changes should unions make in their policy in order to become effective quality improvement participants?
5. What are the things that unions can do on-site to help make organization quality improvement processes succeed?

References

1. UAW president defends policy of cooperation, *Los Angeles Times,* p. 2, June 19, 1989.
2. Rukeyser, L., Frustrated unions take on corporate image—and smear it, *Los Angeles Times,* p. 59, August 24, 1989.
3. United paper workers' corporate campaign news, September 1988.
4. UPIU's Wayne Glenn discusses his union's current goals, *Pulp and Paper,* May 1985.
5. Hoerr, J., The payoff from teamwork, *Business Week,* p. 58, July 10, 1989.
6. Parker, M., and Slaughter, J., *Choosing Sides: Unions and the Team Concept,* South End Press, Boston, p. 14, 1988.
7. Parker, Slaughter, 27.
8. Drucker, P., Workers' Hands Bound by Tradition, *Wall Street Journal,* p. 1, August 2, 1988.

14 Education as the Cornerstone

After Reading This Chapter You Should Know

- The problems with academic training programs for managers.
- Why business schools are reticent to change.
- An idealized curriculum for undergraduate training in business.
- An idealized curriculum for graduate training in management.
- Why classroom style is as important as content in teaching business courses.
- How to teach and encourage team work in the classroom.
- How the customer is critical in encouraging the necessary changes.

Case of the Mission Hammer

Of the three sectors most important to the quality improvement movement—business, unions, and education—the third appears to be the most reluctant to make the changes necessary to an effective contribution. Business-school curricula and teaching styles are outdated and frequently counterproductive. But while this fact has been made obvious by books, magazine articles, and presidential commissions, very little has happened.

One problem is the lack of a hammer. The loss of market share, jobs, profits, and investor and employee confidence is making both corporate and union leadership seek new approaches. Such incentives, however, are generally lacking in academia. It is a world apart, where the refusal to accept reality does not necessarily bring negative consequences, where people who have long since lost their flexibility and desire to adapt retain the power to shape curricula, where leaders enjoy relatively little clout.

Those most affected, the students—unlike employees and stockholders—have no recourse. No grievance committee exists. Even if one does, it is of little use. Undergraduates are too new to effectively evaluate the relevance of what they are learning. MBA students are too busy during their brief two-year stint with course work piled on top of normal eight- to ten-hour days at the office. They grumble, but put in their time, no matter how redundant their studies might be.

Academia's rationalization for ignoring the grumbling is that students do not know what they need, that their input concerning course content would be short-sighted. Such talk, at the graduate level, is unrealistic. The reality of the situation is that colleges and universities have a great number of tenured professors on their faculties who are unwilling to change their ways (sound familiar?). These professors are well paid. They must be used. Their courses, no matter how irrelevant, have to be fit into the curriculum. The president of a large university said recently that perhaps, realistically, this problem did not really make much difference. There was very little that academia could teach top-level executives about management theory and practice.

His statement was a stunning indictment. A great number of top-level executives need help just like everyone else, and the problem is mainly one of education. Such executives have never been encouraged to develop the overview necessary to their new challenge. They have never been trained to think systemically. They continue to focus on pieces rather than on relationships and the whole. They continue to analyze, never learning to synthesize as well.

Doing Things Their Own Way

The fact that some in academia fail to grasp this fact indicates that they have created their own perhaps more comfortable and less demanding reality and that this reality is largely divorced from the reality of the population meant to be served. It is not that the knowledge necessary to encourage the proper perspective in top-level management is lacking. Academicians have been fostering it right along. The contributions of researchers and teachers like Follet, Likert, Herzberg, Maslow, Thorsrud, Emery, Deming, Drucker, Agryris, Schon, Ackoff, Trist, and Bennis can be extremely valuable to executives and future executives.

The problem, therefore, is not a lack of content. Rather, it is the business schools' inability or unwillingness to develop the academic vehicle necessary to present this content properly. The problem, strangely enough, is one of management. Business departments are not being managed effectively. Their leaders are not practicing what they should be preaching.

A major part of this problem, as with top corporate CEOs, has to do with management perspective. Graduate business education has not stayed up with the times. Initially, the MBA was created to give students fresh out of liberal arts colleges with little or no background in business the basics. These basics included technical skills in finance, accounting, production, marketing, human resources, and so on. They allow students to qualify, at least technically, for entry-level management positions.

As time passed, however, both the environment and student profiles changed. Today, most MBA enrollees have already taken introductory business courses during their undergraduate years or as part of corporate in-house training programs. Most have also spent four or five years in the workplace and during this time have developed a great deal of technical expertise. Thus, by the time they return for their MBAs, they are no longer entry-level applicants but middle-level managers and are interested mainly in enhancing their managerial skills.

Unfortunately, most MBA programs have failed to adapt to this new situation. They continue to focus on technical skills, paying only lip service to the managerial side of the coin. Therefore, many of the courses required are, at best, repetitious for students, and, at worst, superfluous.

More specifically, problems stemming from the inability or unwillingness of those managing MBA programs to adapt to changing demands include the following:

1. The fact that MBA programs continue to place too much emphasis on quantitative skills and too little on qualitative or people-related skills that encourage commitment and innovative thinking. Concerning this point, the Business Higher Education Forum says:

> "The rigor applied to financial and quantitative techniques can, and should, be applied to people management. Such courses should include the skills of interviewing, coaching, counseling, negotiating, motivating, and discipline."[1]

Dickerson et al.[2] put it this way:

> "Today the criticism is that business schools are turning out 'number crunchers' rather than managers with good judgment and imaginative ideas . . . It is commonly acknowledged that interpersonal skills are a key competence for managers, but that the contribution of business schools toward the development of such skills is doubtful."

2. The fact that while management training programs at both the undergraduate and graduate levels are good at producing technicians, very few instill the ability to integrate that which is critical to good management. They do not encourage a systemic perspective. If anything, they encourage the opposite. They very rarely talk about how the parts of an effective organization should fit together.

Dickerson et al.[3] cite both the Ford and Carnegie Reports on this issue, discussing the need for:

> " . . . Achievement in terms of developing imaginative, competent, and flexible managers equipped to deal with the unsolved problems of tomorrow . . . Being imaginative in business means having the ability to visualize systemic interconnections among business events and to think counterfactually, that is, to see things not as they are, but as they might be . . . The key decisions in business are non-programmed and often multidisciplinary."

In "The Failure of Business Education—And What to Do About It," Mandt[4] writes:

> "To be blunt, the typical business school curriculum fails to prepare students properly. It fills the student's head with facts—accounting facts, economic facts, marketing facts—and specialized theory, such as "management policy and strategy" or "management behavior and organization theory." But none of this is integrated into any kind of cohesive system."

3. MBAs and the business community have not developed the kind of relationship upon which effective business education should be based. The current role of business leaders as university trustees is mainly to help raise money. It is not to contribute to curriculum development or to offer experience and expertise as a classroom resource. According to Dickerson et al.[5] communication between MBA professors and the business community "appears to be minimal, and perhaps even threatening to both parties."

Time for Change

The most obvious way to deal with the shortcomings defined above is for management program administrators to start listening to the customer and adapting. What they will hear is that a more systemic approach is necessary. At the undergraduate level, this approach should concentrate on developing both the appropriate breadth of perspective and the basic technical skills. Students interested in a management major should be encouraged to take courses in the following business-related areas:

1. Philosophy/logic (traditional tools for improving thought processes).
2. Economics.
3. Science (to develop a better understanding of what is possible technologically).
4. Modern-day ethical and environmental issues.
5. Applied creativity as in painting, writing, and crafts (to foster individuality and the ability to contribute creatively).
6. Public speaking/writing (effective communication is a key to successful management).
7. Psychology.
8. Computing/word processing.
9. Statistics.

Most of these courses are part of a traditional liberal arts core curriculum. In terms of the business major itself, students should be required to take introductory courses in management and organization theory, finance, accounting, marketing, production, and human resources. These courses will provide the necessary frame of reference. Students should then have the opportunity to take one or two additional electives in one or more of these areas. Anything beyond this would probably be a waste. Most corporations prefer to provide the bulk of the technical training required by employees themselves, relating it directly to their own operation.

After the undergraduate degree is earned and students move into the work world, a selection of courses should be offered to give them a chance to further improve technical skills before going on for their MBA. They should be allowed to take as many of these courses in as many areas as desired. One of the advantages of waiting until they are employed to enroll in these advanced specialization courses is that their schoolwork can be tied more closely to their actual job responsibilities. At the same time, students will be more capable of making useful suggestions concerning course content.

In terms of an MBA program, high GMAT scores, especially on the quantitative side, should no longer be the most critical prerequisite to acceptance. The applicant's grade performance at the undergraduate and intermediate levels should be considered, but his or her record as an employee should be given greater weight. At least two recommendations should be required from bosses stating that the applicant is managerial material. This would keep students from wasting time and money when the company has no plans to move them into a higher-level management position. It would also help force performance issues to the surface that had been avoided up to this point.

The MBA program based on systems theory should offer courses on three levels. The first should concern the environment in which the organization must function, both domestic and, increasingly, international. The second should concern the organization as an integrated system. The third should concern interaction between individuals and between members of work groups.

The courses developed to satisfy first-level requirements should offer students an idea of how the organization as a whole fits into the larger environment of which it is a part—the industry, the marketplace, the relevant realms of technology, the community, the nation, and the world. Courses developed to satisfy second-level requirements should give the student an idea of how the organization pieces should fit together, of the systems approach to planning, organization design, communication, decision-making, work design, training, evaluation and reward, and of the systems perspective on integration. Courses developed to meet third-level needs should teach interpersonal, group management, and team building skills—communication, facilitation, counseling, problem-solving, group process, motivation, and so on.

Based on the preceding, MBA-program core courses should include:

1. An introduction to systems theory and the systems approach to management.
2. International business (environmental level).
3. Economic trends and their effects on organization policy (environmental level).
4. Political, legal, and social trends and their effects on organization policy (environmental level).
5. Strategic planning (organization level).
6. Organization structure (organization level).

7. The design and integration of key organization processes including communication, decision-making, work design, training, evaluation and reward (organization level).
8. Team building (work group level).
9. Strategies for making employees more productive (individual level).

Prerequisites to the MBA should be courses ensuring adequate understanding in the key areas of technical expertise. These courses could be taken at the undergraduate level, during the period between undergraduate and graduate study, or at the beginning of the MBA. If students believe that their work experience has provided an adequate conceptual and practical background in one of these areas, a competency examination should be administered and the student exempted if he or she passes it.

And Now for Style

Whereas the problems of undergraduate and graduate management curricula are at least being identified and discussed, those regarding style are receiving almost no attention at all. Reorienting the thinking of bosses who have fought their way up through the ranks and have been doing things the same way for 30 years can be an extremely difficult task. Newer managers tend to be less cynical, less defensive, and more willing to try something new. These, then, are the people who are going to show the way, maybe not tomorrow, but in the next 10 to 20 years. They are also the people that we have sitting in MBA classrooms, reading Ouchi, Peters, Naisbitt, and Pinchot, and discussing the corporate world's current and growing emphasis on quality, on discouraging in-house competition, on encouraging team work, on pushing decision-making down, and on getting employees more involved in shaping their own reality.

While they may be reading these books and discussing these concepts, the young managers are rarely experiencing the new atmosphere being described during their studies. The traditional MBA classroom situation pretty much mirrors the traditional workplace. The relationship between the professor (manager) and students (workers) is adversarial. The professor (manager) makes all the important decisions without input from the students (workers). The students (workers) are responsible to the professor (manager) for completing a certain quantity and quality of work during a predefined period of time. Their reward is based on their production level. A strict hierarchy exists. Bounds of authority are well marked and guarded. Team efforts are relatively rare, and students (workers) are in no way responsible for helping improve their classmates' (co-workers') performance. In fact, they are usually in competition with classmates (co-workers) and can excel only at their expense because the teacher (manager) is grading (paying) on a curve; because only so many A's ($10), B's ($5), and C's ($1) will be awarded.

Thus, no matter what materials are being presented, the traditional MBA classroom situation is reinforcing, by example, the old, competitive, hierarchical, adversarial, mechanistic mode of interaction. It is like a parent who smokes a cigarette while lecturing children on the evils of smoking. It is hypocritical. When professors or MBA program administrators say that the traditional approach to teaching has lasted this long because it is best, we are reminded of the old-time supervisor who makes all the decisions and tells workers not to think, but to just do their jobs: "Because that's the way things have always been done around here."

Academicians should stop reinforcing the old-style values in the classroom and begin introducing those of the quality-seeking workplace. Specifically, they should begin encouraging a more participative atmosphere. They should start sharing their procedural decision-making responsibility and power. The essence of this new atmosphere should be the team approach.

For example, at the beginning of the semester, the MBA course professor could present an outline of topics to be covered along with the basic text and a list of required readings. The professor might then announce that students, from that point on, would run the class. As we have said, a majority of MBA candidates now hold full-time jobs. Therefore, all have undergraduate training, and most have job experience from which to draw. Teams could be formed. Each team could pick a topic or topics, research them, present them, and lead the discussion. These teams, in effect, could be encouraged to function like autonomous work groups, responsible to the manager (professor) for the final results but in control of their own activities.

Teams could be required to provide their own presentation materials. They could be encouraged to bring in actual problems from the workplace around which to build presentations. The professor (manager) could function as a facilitator, encouraging, helping generate requested information, keeping check on team preparations to make sure that members do not wander too far off track, fleshing out presentations when necessary, reviewing and stressing key points, helping coordinate team efforts to avoid overlap, and taking the lead in tying the presentations together into a meaningful whole. This, of course, is exactly the role that new world managers play in companies with successful quality improvement processes.

Just as all members of an autonomous work group receive the same salary and bonus, all members of the presentation team should receive the same grade. The complaint will arise that some students do more work than others. This problem, of course, also exists in the workplace. It is one of the realities that managers must learn to deal with. Again, the team could function like an autonomous work group. At the beginning of the course, the professor could announce that noncontributing members can be dropped from teams if a consensus exists. If the dropped student (worker) is not accepted by another team (work group), the professor (manager) would also have the right to drop (fire) that student (worker) from the class (company).

Presentation grades should be based in part on style—on the creativity of the presentation and on the amount of class participation achieved. This would discourage verbatim readings from the text and other resources while encouraging students to support each other's efforts. Teams could put on plays, sponsor friendly competitions between class members, demonstrate new techniques with the aid of experts, or show homemade videos. Presentation grades should count at least as much as exam grades toward the final grade.

Generally, this approach would tend to raise the class grade average. Yet, emphasis at the MBA level should be on ensuring that students get the most for their time and money, rather than on ranking them. A safe guess would be that a majority of companies are much more interested in the new and enhanced skills and knowledge that employees bring home than in whether they received an A or a B, whether they rank number 1 or number 25. The ideal grading system from this viewpoint would, of course, be a simple pass-fail one. Grades at the MBA level are usually meaningless. Very few students are going on for further degrees. The only purpose of grades is to prove to the sponsoring company that their employees have done the work and learned something. This can be done adequately with a pass-fail system. At the same time, such an approach would further dilute the competitive atmosphere of the traditional classroom situation. It would encourage students to work together as a team, to cooperate, and to help each other.

In summary, then, the advantages of modeling the MBA class after the new world workplace include the following:

1. Students will learn the advantages of a team effort.
2. Rather than trying to beat each other, students will concentrate on learning as much as they can from each other.
3. The classroom experience will help shape positive workplace habits and attitudes.

4. Variety in presentational styles will help make the course more interesting.
5. Students will get a chance to practice presenting themselves and controlling a session.
6. The professor will frequently learn something new and, at the same time, will suffer less chance of burnout.

One final comment is necessary. It concerns undergraduate management programs as opposed to MBA programs. Because students at the undergraduate level frequently lack previous academic training and job experience, it is necessary for professors to exercise more control over both course content and style. However, it is also important to introduce the team approach to learning at this level as well so that students can carry the seeds of change along with them when they graduate and enter the full-time work force.

But What About the Hammer?

A number of ways to improve college and university management education programs have been suggested. The question is, how do we force business department faculty and administrators to budge from their usually defensive posture and pay more attention? Many consider the major problem to be the aforementioned tenure, a system created originally to ensure the academic freedom that professors require for unrestricted inquiry and research, and to make teaching positions more attractive by providing long-term job security. Threats to academic freedom, however, have rarely been a serious issue in this country, at least in modern times, and have almost always been beaten back. At the same time, the long-term security piece has often caused more damage than good. It has allowed too many academicians to function like irresponsible bureaucrats, going through the motions, expanding as little time and energy as possible, accepting as little responsibility as possible, and risking nothing because the reward is assured as long as the boat is not rocked.

The University of California at Berkeley at one point sent shock waves through academia by proposing to scrap its tenure system. Such a move, however, would not be enough. Training is the real issue. College professors spend years, decades immersed in, doing research on, and teaching one approach to reality. Their interpretation of the world is usually selective and narrow. Very few of them, tenured or not, are capable of a systemic perspective concerning curriculum development. Their immediate interest is in protecting their turf, in making sure that their viewpoint receives the attention that it merits. Department chairmen, in turn, are professors who have been promoted. They suffer the same shortcomings that many CEOs possess. They are specialists in economics, English, or chemistry with little training in management skills. As such, they bring to their new jobs a skewed sense of reality that is bound to influence their decisions.

The hammer obviously has to come from outside the department and, owing to the power of the tenured set and the academic freedom issue, from outside the institution. Other colleges and universities are not much help. A comparison of the undergraduate and MBA curricula of the top 30 management departments in the country will show very few differences. Everyone is doing pretty well. The number of applicants continues to be greater than the number of slots to be filled. Why risk change, especially when suggesting even slight adjustments to curricula is bound to spark anguish in the ranks?

The hammer has to come from even further away. It has to come from the customer. Corporations must get fed up with spending large amounts of money to have employees spend long hours learning things that they already know or that are of questionable value. They have to get over the belief that an MBA diploma with the name Northwestern, or Wharton, or Harvard, or Chicago, or Stanford on it automatically qualifies graduates to be superior managers. They have to get over the misconception that an MBA diploma with some college or university name on it is requisite to advancement beyond certain levels of management.

The diplomas, and the names of institutions of higher learning on them, in many cases, are little more than impression pieces, because the substance supporting them is the wrong substance. Corporations, unions, and other organizations have to start paying closer attention to what employees are actually studying. When they do, and realize, or admit, that, despite the growing body of evidence, the growing chorus of pleas, students are still not being prepared for modern-day management positions, they will do the obvious—develop their own comprehensive management training programs.

One large corporation, a group of small ones, or a corporation-union partnership will develop a curriculum and hire hand-picked professors to come in and teach. Despite the fact that they will pay these people well, they will end up spending far less for employee education. At the same time, they will be able to constantly adjust and adapt their curriculum, learning from both students and the environment. They will award some type of degree. It will not be recognized by academia, but will quickly prove its value in terms of content and, therefore, will rapidly gain legitimacy and popularity.

Academia will worry loudly about the business sector not having the proper perspective. It might even try to blackball professors willing to teach in these renegade schools. In order to remain competitive, however, MBA programs will finally be forced to begin listening seriously and to react in the right manner.

Once the dust settles, the end result will most likely be that instead of the business sector surrendering employees to the academic sector to be educated as academia sees fit, the two sectors will work together, pooling expertise and resources to ensure that students receive the most relevant education possible. The end result will be that management professors will become more directly involved in the real world of business that they are helping to mold, and that corporate executives and union officials will get over their uneasiness around academicians and make better use of their talents. The end result, in sum, will be a win-win, quality-seeking situation from which all sectors benefit and learn.

Topics for Discussion

1. What is the best vehicle or combination of vehicles for providing managers with the education they need?
2. What are the obstructions to the development of these vehicles and how can they be dealt with?
3. What are the subjects that need to be covered in formal management education?
4. How can the business and education sectors better support each other?
5. What are the ways of encouraging a cooperative atmosphere in the classroom?
6. How would you, as students, teach this course?

References

1. America's business schools, priorities for change, a report by the Business-Higher Education Forum, Washington, D.C., May 1985, p. 3.
2. Dickinson, R., Herbst, A., and O'Shaughnessy, J., What are business schools doing for business? *Bus. Horiz.*, November-December 1983, p. 46.
3. Dickerson, Herbst, O'Shaughnessy, 47

4. Mandt, E. J., The failure of business education—and what to do about it, *Management,* p 48, August 1982.
5. Dickerson, Herbst, O'Shaughnessy, 51

15 One More Time, the Systems Approach: The Good Shepherd Effort

After Reading This Chapter You Should Know

- What the Joint Commission on Accreditation of Healthcare Organizations (JCAHO) philosophy concerning organization improvement has in common with the systems approach philosophy.
- What the JCAHO quality improvement model has in common with the Baldrige model.
- Why Good Shepherd chose the systems approach to quality improvement.
- What it takes to start a systemic quality improvement effort in the health care sector.
- How a quality improvement effort can be kept from burying an organization.
- Ways to reward quality improvement in the nonprofit sector.
- What the primary focus of all quality improvement efforts (mentioned in the preface) must be.

JCAHO Follows the Baldrige Award Lead

The health care sector is currently experiencing a great deal of turbulence. Government policies regulating it, both on national and state levels, are in a state of flux. Changes in Medicare and other medical insurance policies are creating the need for sharp cutbacks. JCAHO, in its efforts to make management more effective, is encouraging the team approach. It wants employees other than top-level managers to become more involved in decision-making and process improvement.

JCAHO talks systemically. For example, in section P1.1 of the organization's guidelines we find it saying that, "Performance-improving activities are most effective when they are planned, systemic, and organization-wide, and when all appropriate individuals and professionals work collaboratively to implement them. Too often performance-improvement efforts are isolated within specific departments, units, or professions." Then the document goes into great detail describing action steps for improvement efforts, focusing on the introduction of appropriate performance measures. Again, in the section on "Standards, Intents, and Examples for Improvement," we read, "The departments and disciplines that have an impact on a process (should) participate in the decision to test or implement an improvement strategy, and (should) select performance measures against which to assess the action taken."

But, then, in section P1.5 the guidelines revert to the norm. They say that, "*Hospitals leaders* (should) set improvement priorities," and a little further on that, "The *leaders* (should) then implement a hospital-wide approach to improve targeted processes." As we read we learn that design teams are formed around individual projects. They will do the work, then report back to the process leaders. We learn that JCAHO mandates the approval and prioritization of projects by senior-level management to ensure that the projects fit within the frame of overall organization objectives and to ensure the most efficient use of resources.

This is basically the Baldrige model, top-down, built around what systems people call task forces. It has all the problems of the top-down, task force driven model. These include the fact that such efforts rarely are able to make effective use of the expertise of lower-level employees; that the important decisions are made by senior-level management, not by the people most familiar with the problem. It includes the fact that, especially in the current environment, top-level management is already overloaded, that this is another responsibility that frequently gets put on a back burner. It includes the fact that the ground rules are not in place to help protect task forces from turf battles when information is needed or when their project affects others. It includes the fact that a flotilla of task forces pursuing their own objectives does not add up to an integrated, systemic effort that develops the necessary understanding and ownership in employees. It includes the fact that training is done up-front and that what is being put into place is too complex, takes too much time to learn, and that the desired improvement are a long time in coming, if ever.

As a result, a great number of health care organizations are looking for an alternative approach, for ways to improve their efforts. The have no choice. In order to survive, they must find ways to cut costs, to become more productive, and they must do so quickly. Good Shepherd Home, a rehabilitation hospital in Allentown, PA with a long-term care facility and a work service component, found itself in this position in 1992. Owing to increasing marketplace pressures, Good Shepherd began losing money. To remedy this situation, the Home hired a traditional, highly respected, expensive quality consulting firm, once again Philip Crosby's, to teach employees to work more efficiently. Crosby's approach, as always, was top-down and training-driven. Almost all of the organization's 850 employees were required to attend classes on topics including the cost of missing quality, team building, and the effective use of various statistical measurement tools. Top management was repeatedly warned by the consultants doing the training that the desired change would take time.

Meanwhile, in March of 1993, because Good Shepherd was still operating in the red, the Home saw no alternative other than to downsize. Good Shepherd released 83 employees. The effects of this decision were immediate. Morale plummeted. A period of mourning set in. Good Shepherd had traditionally been run like a family with upper-level management (all male) playing the role of the benevolent father. Job security had not normally been an issue. Now, it appeared that top management could no longer be trusted. The foremost question in everyone's mind immediately became, "Who's next?" Also, the work-load increased. Those remaining had to take on additional responsibilities without adequate training. The quality process disappeared in the confusion. It was no longer seen as a solution, but rather as another responsibility piled onto an already too full plate. And the bottom line remained in the red, so that the threat of another round of downsizing was very real.

At this point Frank Pyne, Good Shepherd's CEO, became aware of the systems approach to organization change and decided to try it. The systems approach appealed to Mr. Pyne because it centered on immediate and real employee empowerment. He believed that the only way to rescue his organization from its downward spiral and to regain employee confidence was to truly involve the work force in the effort. The systems approach, more than any other he was familiar with, seemed to encourage this, and to get things going quickly.

During his introduction of this next round in Good Shepherd's efforts to reverse its fortunes Mr. Pyne attempted to deal with the cynicism employees were no longer hiding, and, at the same time, with the related issue of job security. He said he knew what employees were thinking. They were thinking, "Here we go again. More training. More charts and graphs to fill out. More responsibilities." He asked simply that they give this new approach a chance. Concerning further downsizing, which he knew was also on everyone's mind, Mr. Pyne said that he would do everything possible to avoid another round, but that he needed everyone's help.

Get Set, GO!

The familiarization phase started immediately at Good Shepherd. Five half hour sessions were scheduled during a ten day period to introduce the new model to all levels of staff. The sessions were run by a consultant trained in the systems approach and the new quality process head facilitator, Mary Ellen Place, who had previously been Director of Social Work Services. The systems approach and the five-phase model were briefly explained, with emphasis on team building. The major point made was that the employees were now in charge. If the new process did not produce the desired results, if it did not improve their work lives, they could change it, or shut it down. They were being given the power to do so.

The second phase, team building, began the following week. The head facilitator and consultant worked together to map out the network of teams that would be formed, roughly one per department. Some teams, however, were defined, instead, by function. The hospital secretaries, for example, had their own team in order to deal with their own secretarial issues before getting involved with any others. Maintenance also had its own team.

These initial quality action teams (QATs), as hospital staff called them, were given the traditional systems approach ground rules. They included only hourly and professional staff. The reason for not allowing managers to sit in on them was to give the workers a true feeling of empowerment, to allow them to talk openly about concerns without having to worry about their bosses' reaction.

Potential team facilitators were identified by the head facilitator, and the first team was brought up by the consultant with the head facilitator observing. All facilitators from that point on learned their skills by serving initially as apprentices. The head facilitator at Good Shepherd learned to bring up teams by watching the consultant do it, then by utilizing what she had learned with the consultant in support. The team facilitators learned by first watching the head facilitator or consultant run their teams' meetings, then taking over with someone in support. Eventually, the facilitators developed their own support network which met monthly. Facilitators identified their own ongoing training needs, developed and frequently delivered their own training programs, some of them in the classroom.

The first QAT, Occupational Therapy, was working on its initial project within three weeks of formation and had generated and prioritized a list of some 27 other issues it wanted to address. Within 6 months 24 QATs had been organized, spanning the organization, integrated by the facilitator network. Each had its own list of 15 to 40 projects to work on, beginning with improvements to its immediate operation, then moving outward to address relationships with other functions.

The second type of team formed was the management team. Managers in individual departments began meeting weekly to identify and to work on projects they were interested in. The issues they addressed were usually broader than those addressed by the QATs. For example, the clinical managers focused on designing an interdisciplinary approach to patient/family education. The managerial teams followed the same ground rules as QATs and had facilitators.

The third and final type of team formed was the project-specific task force. A task force could be formed by management around any project not being addressed by a QAT or by a management team. Task forces could also be formed by individual QATs or management teams to deal with large, long-term problems involving a number of departments. Task forces lasted only until the project they were built around ended, then disbanded. Task forces had a facilitator, usually the head facilitator, and were supported by the same ground rules as the other teams. An example of a project assigned to a task force was the design of a staff back injury prevention program in the long-term care division.

Eventually, after the consultant had left, one other group was formed. This was the quality improvement team (QIT), an interdisciplinary, multidivisional team that continued the familiarization phase and provided answers for employees and outsiders. It also continuously reviewed the overall process looking for ways to improve it. Finally, the QIT took the lead in developing systemic (comprehensive) benchmarking techniques, selecting appropriate targets for benchmarking, and carrying out the benchmarking process.

Once the Good Shepherd work force realized that it really had been empowered to identify issues, generate solutions, and to orchestrate improvements, buy-in came rapidly. Teams began identifying their own job-related training needs. Basic statistical measurement techniques were eventually made available to facilitators, who were instructed to introduce them when team members had identified a project situation where they might be of use.

Finally, Good Shepherd, under the leadership of Frank Pyne, is working to introduce interactive planning, a systemic planning paradigm that fits neatly with the systemic quality improvement model. The planning phase of quality improvement processes usually takes the longest to get started because the models most companies use are not systemic. Companies, therefore, have trouble making a connection between what they are doing in terms of planning and comprehensive quality improvement. They do not understand the interdependencies.

In terms of further refinements, the process continues to evolve and to develop new challenges. One, which is not so new, concerns a group of long-term, older managers similar to that found in most organizations (professors in academia), the managers Tom Peters talks about, who continue to have trouble accepting this new way of doing things, despite obvious operational and bottom line benefits. To the credit of these people at Good Shepherd, they have not tried to block the process, despite their misgivings. It is hoped that the interactive planning effort, in which they will play the key role, will help gain their active support. It is also understood that some of them will never be able to accept what is going on. It is counter-intuitive to their conditioning. They cannot help but see it as a threat. They are too far along, have made too many commitments, and too many sacrifices to change. They are frozen into their position and mindset and simply want to hang on long enough to retire. Such people cannot be blamed for their attitude. At the same time, however, they cannot be allowed to block the necessary changes.

In terms of a reward system that ties everyone's take to bottom line improvements, this can be done in nonprofit as well as for-profit organizations. As the bottom line increases owing to a combination of better marketing, better service, and increased efficiencies, bonuses can be awarded. Again, they should not be awarded for the best project or to the best team. This breeds competition when organization-wide cooperation is key to success. Rather, to be most effective, the bonus should be shared equally organization-wide. Other forms of reward are also popular. The most obvious in the nonprofit sector is time off from the job. Quite frequently this reward is as appreciated as or even more appreciated than a monetary one. And, again, if lower-level rewards such as a write-up in the organization newspaper, a certificate, lunch with the CEO, are given out based on team performance, it is better that all teams are able to earn them upon reaching a certain level of accomplishment, rather than just the "winner." Although friendly competition or co-optition (a new buzzword) at this level is not necessarily detrimental to the process, it stills sends the wrong message.

Simplicity Is Best: Lessons Learned

Lessons learned by Good Shepherd in a relatively short period of time once the systems approach to quality improvement had been adopted include the following:

Lesson number 1 It does not take a lot of time or money to start a comprehensive quality improvement effort and to begin producing positive results. It does not take months of preparation, hoards of consultants, lectures, two-inch thick notebooks, charts, strobe lights, or pep rallies. It takes one person, either in-house or a consultant, who understands the ingredients and interactions necessary to a common-sense, comprehensive (systemic) model and who is willing to share his or her knowledge with others. In addition, it takes a staff willing to learn. This, however, is rarely a problem, especially at the lower levels.

Lesson number 2 It also takes top-level executives who are willing to go all the way. These are people like Ralph Stayers (Johnsonville Foods), Richard Teerlink (Harley Davidson), Richard Semlar (Semco), Mike Turk (ATT Reading), W.L. Gore, Bob Goins, and Frank Pyne, the small but growing group of leaders willing to fully empower their work force, to step back and let others, many others, take the lead. Most CEOs, as we have said, want increased employee involvement in their improvement efforts, but they also want management to stay in charge. They see no other alternative. Their public reasoning is that management's overriding perspective is necessary to keep the process on track, that if they do fully empower the work force and let employees identify and run their own projects, time and resources will be wasted. Their gut reasoning, again as we have said, has more to do with ego than with logic. They have come too far, fought too hard in a usually cutthroat environment to give up the power they now enjoy.

Lesson number 3 Employees learn more quickly by doing than by listening. Training large numbers of people in a classroom setting in process-related skills up-front usually wastes both time and money. First, many of them already have the requisite problem identification, problem solving, conflict resolution, and teaming skills. They use them every day on the job and in their private lives. Second, when learning in the classroom, especially up-front, participants lack a real-life frame of reference to relate what they are learning to. Give them the opportunity to contribute, without the frills, and they will do so, enthusiastically. The apprenticeship approach to training is the oldest in existence. It has survived this long because it is also the most effective. It makes a good fit with systemic quality improvement efforts. In order for it to be implemented effectively, however, human resources departments will have to break with tradition.

Lesson number 4 Do not bury your organization under a quality improvement process. It should not interfere with normal job responsibilities. After the start-up, a lot of time does not need to be spent in team meetings. The main purpose of team meetings is to review project action steps completed during the preceding week and to identify new ones. This work usually does not take more than one half to three fourths of an hour if the facilitator is doing his or her job correctly. Also, no employee should sit on more than one functionally defined quality team and one task force at the same time. We know of companies where employees sit on as many as 12. This is a mistake. It defeats the purpose of the exercise by taking people away from their primary operational duties. Because, in the systems model, the quality team network is organization-wide and integrated by the ground rules, everyone affected does not need to be involved in every project from start to completion.

Lesson number 5 Good Shepherd is still in the process of learning that an interactive (systemic) planning effort is a necessary ingredient of any quality improvement effort that hopes to be successful in the long run. Interactive planning provides the overall framework of organization objectives for teams into which to fit their projects. Because it is participative, organization-wide, and ongoing, it becomes a primary integrating mechanism for the process. Because it is led by upper-level management, it gives these people the final say concerning the direction of the process, but without having to control the activities of quality teams, management teams, and task forces.

In summation, Good Shepherd, like International Paper, learned the hard way that simplicity is best when mounting a quality improvement process. Once the lesson was learned, however, and the systems approach was discovered, Good Shepherd's effort, like those of the two International Paper mills, began gaining momentum. The necessary primary focus, the one we mentioned in the Preface as being the one around which everything else revolves, was, in both cases, the true empowerment of employees on all levels. Both companies learned to trust their employees enough to allow them to lead the change effort in their own areas of expertise, and that made the difference. At the same time, however, both companies realized the need to integrate employee efforts and to make sure that they fed into the overall mission of the organization. This was done by building an organization-wide network of teams whose efforts were integrated by facilitators, by starting an interactive planning process, and by introducing ground rules agreed to by all.

The Good Shepherd quality process is currently four years old. The results speak for themselves. The team network at this point has completed approximately 406 projects. It has addressed, amongst other things, work design issues, improvements in technology and management systems, improvements in the work environment, community relations, customer satisfaction, office design, satellite functions, and relations with other health care institutions. What employees have learned through the process has become part of the culture, changing the way decisions are made on a day-to-day basis. Finally, the process has contributed to a general improvement in morale, despite the uncertainty that currently pervades the health care industry.

Topics for Discussion

1. Why does the health care industry tend to favor the Baldrige model over the systems model?
2. What differences exist between the systems model used at Good Shepherd and the systems model used at the two International Paper mills?
3. Compare the start-up of your organization's quality improvement process with that of Good Shepherd.
4. Discuss the differences found in for-profit and nonprofit organization quality process-related reward systems.
5. And, one last time, discuss why real employee empowerment must be the focus of systemic quality improvement processes.

Bibliography

Ackoff, R. L., *Creating the Corporate Future,* John Wiley & Sons, New York, 1981.

———, Does quality of life have to be quantified?, *Gen. Syst.,* Vol. 20, 1975, pp. 5–15.

———, *Management in Small Doses,* John Wiley & Sons, New York, 1986.

———, The corporate rain dance, *The Wharton Magazine,* Winter, 1977, pp. 36–41.

———, *The Democratic Corporation,* Oxford University Press, New York, 1997.

———, *Redesigning the Future,* John Wiley & Sons, New York, 1986.

Ackoff, R. and Emery, F., *On Purposeful Systems,* Aldine-Atherton, New York, 1972.

Adler, M., *Aristotle for Everybody,* Macmillan, New York, 1978.

Allio, R., Executive retraining: the obsolete MBA, *Bus. Soc. Rev.,* Summer, 1984.

Anon., America's business schools: priorities for change: a report by the Business-Higher Education Forum, Washington, D.C., May 1985.

Avedesian, J., Cowin, R., Ferguson, D., and Roth, W., Beyond crisis management, *Pulp Pap. Inte.,* February, 1986, pp. 50–53.

Batter, W., Productivity and the working environment, *The Wharton School of the University of Pennsylvania Lecture Series,* March 17, 1985.

Bennett, A., Business takes out its trimming shears, *Wall Street Journal,* p. 1, October 5, 1989.

Castro, J., Where did the gung-ho go?, *Time,* p. 53, September 11, 1989.

Chems, A. and Davis, L., *The Quality of Working Life,* Vols. 1 and 2, Collier Macmillan, London, 1973.

Anon., Company courses go collegiate, *Bus. Week,* pp. 90–92, February 26, 1979.

Anon., The corporate elite, *Bus. Week,* p. 28, October 21, 1989.

Anon., Developing managers not a corporate priority, *Wall Street Journal,* p. 1, April 18, 1988.

Dickinson, R., Herbst, A., and O'Shaughnessy, J., What are business schools doing for business?, *Bus. Horiz.,* November–December, 1983, pp. 46–51.

Drucker, P., Workers' hands tied by tradition, *Wall Street Journal,* p. 2, August 2, 1988.

Emery, F. and Thorsrud, E., *Democracy at Work,* Martinus Nijhoff, Leiden, 1976.

George, C., *The History of Management Thought,* Prentice-Hall, Englewood Cliffs, NJ, 1986.

Gharajedaghi, J., *Interactive Redesign: Third Generation Systems Thinking,* Quality Press, Chico, CA, 1998.

———, *Toward a Systems Theory of Organization,* Intersystems Publications, Seaside, CA, 1985.

Hackman, R. and Suttle, L., *Improving Life at Work,* Goodyear Publishing, Santa Monica, CA, 1977.

Hirose, K., Corporate thinking in Japan and the U.S., *Jpn. Econ. Found. J. Jpn. Trade Ind.,* May, 1989, pp. 23–24.

Hoerr, J., Is teamwork a management plot? Mostly not, *Bus. Week,* February 20, 1989, p. 56.

———, The payoff from teamwork, *Bus. Week,* July 10, 1989, p. 58.

———, Power-sharing between management and labor: it's slow going, *Bus. Week,* February 17, 1986, p. 50.

Hofstadter, R., *Social Darwinism in American Thought,* George Braziller, New York, 1959.

Hymowitz, C., Employers take over where schools fail to teach the basics, *Wall Street Journal,* p. 1, January 22, 1981.

Anon., Is the boss getting paid too much?, *Bus. Week,* pp. 46–48, May 1, 1989.

Anon., *The Japanese Economic Journal,* pp. 34–39, December 20, 1986.

Anon., *Japanese Times,* p. 4, December 31, 1987.

Jardim, A., *The First Henry Ford: A Study in Personality and Business Leadership,* MIT Press, Cambridge, MA, 1970.

Jenkins, D., *Job Power,* Penguin Books, Baltimore, 1973.

Johnston, J. and Associates, *Education Managers,* Jossey-Bass, San Francisco, 1986.

Kadar, S., *Frederick Taylor: A Study in Personality and Innovation,* MIT Press, Cambridge, MA, 1970.

Kamber, I., Thank labor for our high quality of life, *Los Angeles Times,* p. 15, September 4, 1989.

Kreitner, R., *Management,* 3rd ed., Houghton Mifflin, Boston, 1986.

Lefton, R. and Buzzotta, V., Teams and teamwork: a study of executive-level teams, *Nat. Prod. Rev.,* Winter, 1987–88, pp. 23–25.

Lewis, P., Goodman, S. and Fandt, P., *Management: Challenges in the 21st Century,* West Publishing, New York, 1995.

Likert, R., *New Patterns of Management,* McGraw-Hill, New York, 1961.

Mandt, E., The Failure of Business Education—and What to Do About It, *Management,* August, 1982, pp. 47–50.

Maslow, A., *Motivation and Personality,* 2nd ed., Harper & Row, New York, 1970.

Moskal, B., The sun also rises on GM, *Industry Week,* September 5, 1988, pp. 100–102.

Moskal, B., Born to be real, *Industry Week,* August 2, 1993, pp. 14–18.

Naisbitt, J., *Megatrends,* Werner Books, New York, 1982.

Anon., *Nikkei New Bulletin,* April 4, 1986.

Park, E., It's revolutionary! And it works!, *AT&T Journal,* June 1987, pp. 4–6.

Parker, M. and Slaughter, J., *Choosing Sides: Unions and the Team Concept,* South End Press, Boston, 1988.

Place, M. E. and Roth, W., Simplicity is best: the learning experience of Good Shepherd Center, *J. Health. Qual.,* May/June, 1998, pp. 20–24.

Renwick, P. and Lawler, E., What you really want from your job, *Psychol. Today,* May, 1978, pp. 53–65.

Roth, W., A new role for unions, *J. Qual. Partic.*, September, 1990.
———, Are safety programs really effective? *Pulp Pap. Int.*, March, 1988, pp. 50–51.
———, Dos and don't of quality improvement, *Qual. Prog.*, August, 1990, pp. 85–90.
———, Five phases to success, *J. Qual. Partic.*, June, 1989, pp. 26–32.
———, Get training out of the classroom, *Qual. Prog.*, May, 1989, pp. 62–65.
———, Middle management: the missing link, *TQM Mag.*, Vol. 10, No. 1, 1998, pp. 6–10.
———, *Problem Solving for Results,* St. Lucie Press, Delray Beach, FL, 1998.
———, Quality: an opportunity for human resources, *Total Qual. Rev.*, May/June, 1994, pp. 7–10.
———, Quality: rebirth of the systems approach, *Qual. Dig.*, January, 1991, pp. 52–57.
———, The dangerous ploy of downsizing, *Bus. Forum,* Fall 1993, pp. 5–8.
———, *The Evolution of Management Theory: Past, Present, Future,* St. Lucie Press, Delray Beach, FL, 1993.
———, The slippery slope, *TQM Mag.*, November–December, 1993, pp. 43–47.
———, The view from the top and successful quality processes, *J. Qual. Partic.*, September, 1994, pp. 46–49.
———, Today's MBA: a lot to learn, *Personnel,* May, 1989, pp. 46–51.
———, Try some quality process glue, *J. Qual. Partic.*, December, 1989, pp. 20–27.
———, What's going on down in Louisiana?, *Pulp Pap. Int.*, September, 1987, pp. 87–89.
Rukeyser, L., Frustrated unions take on corporate image—and smear it, *Los Angeles Times,* p. 59, August 24, 1989.
Schon, D., *Beyond the Stable State,* W. W. Norton & Co., New York, 1971.
Semler, R., Managing without managers, *Harv. Bus. Review,* September–October, 1989, pp. 76–84.
Slutsker, G., Hog wild, *Forbes,* pp. 45–46, May 24, 1993.
Stayer, R., How I learned to let my workers lead, *Harv. Bus. Review,* November–December, 1990, pp. 66–83.
Stayer, R., Managing The Journey, Inc., pp. 47–52, November, 1990.
Taylor, F., *The Principles of Scientific Management,* Harper, New York, 1911.
Anon., "This is the Answer," *Forbes,* July 5, 1982, pp. 56–58.
Trist, E., The evolution of socio-technical systems, in *Issues in the Quality of Working Life,* Vol. 2, Ontario Ministry of Labor, Ontario, 1980.
Trist, E., Higgins, G. W., Murray, H., and Pollock, A. B., *Organizational Choice,* Tavistock Institution Publications, London, 1963.
Anon., UAW president defends policy of cooperation, *Los Angeles Times,* p. 2, June 19, 1989.
Anon., United Paper Workers Corporate Campaign News, September, 1988, p. 1.
Anon., UPIU's Wayne Glenn discusses his union's current goals, *Pulp Pap.,* May, 1985, p. 17.
Anon., The vision of MITI policies in the 1980s, Ministry of International Trade and Industry, Tokyo, 1979.
Anon., When companies tell business schools what to teach, *Bus. Week,* pp. 60–61, February 10, 1986.
Yardley, J., Should universities remain shelters for the slothful? *Los Angeles Times,* p. 12, January 11, 1990.
Whitsett, D. and Yorks, L., *From Management Theory to Business Sense: The Myths and Realities of People at Work,* AMACOM, New York, 1983.

Index

D

E

N

O

P

Q

R

U

V

W

X